SpringerBriefs in Computer Science

SpringerBriefs present concise summaries of cutting-edge research and practical applications across a wide spectrum of fields. Featuring compact volumes of 50 to 125 pages, the series covers a range of content from professional to academic.

Typical topics might include:

- A timely report of state-of-the art analytical techniques
- A bridge between new research results, as published in journal articles, and a contextual literature review
- A snapshot of a hot or emerging topic
- An in-depth case study or clinical example
- A presentation of core concepts that students must understand in order to make independent contributions

Briefs allow authors to present their ideas and readers to absorb them with minimal time investment. Briefs will be published as part of Springer's eBook collection, with millions of users worldwide. In addition, Briefs will be available for individual print and electronic purchase. Briefs are characterized by fast, global electronic dissemination, standard publishing contracts, easy-to-use manuscript preparation and formatting guidelines, and expedited production schedules. We aim for publication 8–12 weeks after acceptance. Both solicited and unsolicited manuscripts are considered for publication in this series.

**Indexing: This series is indexed in Scopus, Ei-Compendex, and zbMATH **

Robert Kudelić

Feedback Arc Set

A History of the Problem and Algorithms

 Springer

Robert Kudelić
Faculty of Organization and Information
Science, Theoretical and Applied
Foundations of Information Science
University of Zagreb
Varaždin, Croatia

ISSN 2191-5768　　　　　　　ISSN 2191-5776　(electronic)
SpringerBriefs in Computer Science
ISBN 978-3-031-10514-2　　　　ISBN 978-3-031-10515-9　(eBook)
https://doi.org/10.1007/978-3-031-10515-9

This Springer imprint is published by the registered company Springer Nature Switzerland AG
The registered company address is: Gewerbestrasse 11, 6330 Cham, Switzerland

Foreword

The feedback arc set problem is a one of the quintessential problems of algorithmics and, more generally, of computer science. The problem is easily stated: for a given directed graph $G = (V, E)$, find the smallest subset $E' \subset E$ such that $G' = (V, E \setminus E')$ is acyclic. The problem is well-known and has been investigated by the best minds in computer science. It arises as a fundamental issue in many problems in industry, from circuit design to logistics. Yet, the problem has proven very difficult to solve.

Over the last fifty years, methods to solve the feedback arc set problem have spanned the length and breadth of the techniques of computer science: paradigms of graph algorithms, well-designed data structures, optimization via integer programming, heuristics with proven performance guarantees, "soft computing" methods such as ant algorithms, and stochastic methods such as Monte-Carlo algorithms.

The author Robert Kudelić has published papers on several disparate approaches to the feedback arc set problem. His book takes the reader chronologically through the feedback arc set literature, and as such it provides a historical compendium of techniques in computer science.

Sydney, NSW, Australia   Peter Eades
January 2022

Preface

As is true for any subject in science, there comes a time when one needs to look back, and take a gaze upon work done thus far. And if we are to choose such a time for a classical computer science problem known as feedback arc set (FAS), and more than half a century has passed since its fledgling days, this is a good time to work on that gaze. A general introduction to the problem will be presented, hardness and importance will be argued. Span will be given from the origin and the name, through problem definition and various versions, all the way to the dual version and a link with the feedback vertex set and the forcing problem—practical relevance will be covered as well.

By building on that foundation, a review of the algorithms will be given, with special emphasis on FAS. The goal of this review is to give both depth, through covering algorithms, and breath, through spanning lifetime of the FAS problem. It is the intention that, through this book, any interested party will be able to quickly find relevant information, fostering both learning and research.[1]

There are a number of people that were of help to me during preparation of this book, and therefore a few words of gratitude are in order. I would like to thank my college for suggesting that the material which is found in this work, a work in progress at that time, would be a good fit for a book—the idea quickly grew in my mind, and the publisher was soon found.

I would also like to express my sincere gratitude to Peter Eades, well known for his expertise in graph problems, for being so kind and taking the time to write the foreword to this book.

---

[1] Online supplementary material is available at: https://cs.foi.hr/fas/book/.

A special appreciation would go to my father for enabling my academic career. Gratitude is in order for a colleague of mine, Nikola Ivković, who was very helpful in acquiring a number of papers for me. And I can't forget Wayne Wheeler, for the support and for being such a nice person to talk too, it was wonderful.

Varaždin, Croatia   Robert Kudelić
April 2022

Contents

Part I
Overview of Findings

Chapter 1
Feedback Arc Set

1.1 The Problem and the Name

In this work we will present a chronology of algorithms for a well known classic computer science problem known under a name Feedback Arc Set (FAS).

The problem can sometimes be found under a guise of Quadratic Assignment (QA) problem, since it can be formulated as such (with QA being a more general problem), and then solved by algorithms for the QA [48]. At other times FAS can be found under a terminology of Linear Arrangement (LA), or sequencing, or of the Median Order [13, 14, 28]. It is also possible to find it under a name Hitting Cycle (HC) problem, since one has to hit every cycle in a collection of cycles [22]. Sometimes it is possible to find it under a name of Minimum Dominating Set problem [34, 67].

But in almost all cases a name labeled, and through the years accepted, to a problem of finding arcs with which removal graph becomes acyclic is Feedback Arc Set, or MFAS if minimum is what one wants to find—in its dual the problem is called Maximum Acyclic Subgraph (MAS), or sometimes Maximum Consistent Set [14]. As it can be gathered from the literature, it seems that the name was entrenched by Younger in his 1963 paper on MFAS and directed graph [66].

1.2 Origin

The problem has its origins in the analysis of sequential switching circuits with feedback paths [66]. Runyon originally suggested it after he observed that such analysis would be simplified if one knew a set which represented minimum feedback

Fig. 1.1 Feedback Arc Set
problem illustration

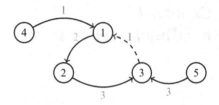

arc set[1] [4, 39, 61, 66]. Since then, because of its importance in feedback systems, MFAS has attracted an ever increasing attention.

FAS was one of Karp's core 21 NP-Complete problems [41], and is long known to the scientific community. A few decades have passed, and many algorithms have been proposed, proved, and developed. There is now quite a forest of possible avenues of solving FAS, and a review is needed—such a review will help anyone who is entering the field, and wants to give its own contribution—not only as a reference for the state of the art but also for a quick comparison with the state of the art, or as a material quickly pointing in the right direction. Such a review will also be of help to anyone who wants to get a gist of the FAS, and it will also be valuable as a reference material for everything important FAS.

1.3   Description

When we are speaking about FAS we are thinking in terms of breaking a number of arcs (in the literature, symbols A and E are often used interchangeably) with which removal the graph in question will become acyclic [26]. In its optimization version where we seek a minimum, the problem is known as MFAS [6, 20, 46, 66]. By speaking about FAS we are speaking about decision version of the problem that is stated as follows.

Instance:	Directed graph $G = (V, A)$, a positive integer $K \leq |A|$.
Solution:	Is there a subset $A' \subseteq A$ with $|A'| \leq K$ such that A' contains at least one arc from every directed cycle in G?

Previous definition for Feedback Arc Set in a decision version was taken from [26]. Graphical illustration is presented in Fig. 1.1. In its optimization version the problem is stated as follows.

---

[1] According to [66] it was mentioned in [61] in the appendix (among a list of research problems).

Instance: Directed graph $G = (V, A)$, a positive integer $K \leq |A|$.

Solution: Is there a subset $A' \subseteq A$ with $|A'| \leq K$ such that A' contains
minimal number of arcs with which to break
every directed cycle in G?

Considering that there is always some cost for finding, and removing an arc, or reversing it, it is often the case that one wants to make a digraph acyclic by removing minimal number of arcs. It is also sometimes necessary to invert, or remove, minimal number of arcs, for the reason of a problem one wants to solve, and then of course one wants to achieve it in a fastest way possible.

"In any directed graph, the size of minimum feedback arc set is at least the number of maximum arc disjoint directed cycles in the directed graph" [31]—with the minimum feedback arc set being equal to the number of maximum arc disjoint directed cycles on planar digraph [49].

1.4   Linear Programming Formulation

Given graph $G = (V, E)$, with arc weights (G, w), and the set C with all cycles in graph G, minimum Feedback Arc Set problem with arc weights can then be formulated as the following integer programming problem [22]

$$
\text{(MFAS)} \begin{cases} \min \sum_{e \in E(G)} w(e) x_e \\ \text{s.t.} \sum_{e \in \Gamma} x_e \geq 1 \quad \forall \Gamma \in C \\ x_e \in \{0, 1\} \quad \forall e \in E(G) \end{cases} \tag{1.1}
$$

where Γ represents a cycle in C [22]. If the problem is relaxed, constraints for x_e are replaced with $x_e \geq 0, \forall e \in E(G)$, thus obtaining a fractional FAS [22].

Considering that problem of FAS is a covering problem, its (linear programming) dual is a packing problem [22]. "In the case of the feedback arc set problem this means assigning a dual variable to all interesting cycles to be hit in the given graph, such that for each arc the sum of the variables corresponding to the interesting cycles passing through that arc is at most the weight of the arc itself." [22]

1.5   Inherently Hard to Solve

Unfortunately FAS, and MFAS naturally, is one of those problems computers are having difficulty to solve efficiently, one of those problems that are simply refusing to obey. It is known that FAS is NP-Complete [41], even for a class of digraph where input and output degree of a vertex does not exceed 3 [31], this has been as such for a long time now. And if we are more specifically interested in its optimization version, it is well known as well that MFAS is particularly difficult to solve, and is in fact NP-Hard [48]. It should be noted as well, that the problem of Feedback Arc Set stays NP-Complete for "line digraphs, even when every clique has at most size three" [27, 35]. If one is to be concerned with counting, then computing the number of minimal feedback arc sets of a directed graph is #P-hard [60].

These facts of themselves are strong proof of the difficulty of the problem, although that difficulty does not end here—MFAS is also APX-Hard as well, and as such has a hard limit on its approximation quality [40]—its approximation constant stems from a well known problem named Vertex Cover, through structure preserving reduction, and amounts to $c = 10\sqrt{5} - 21 \approx 1.36067$ [6, 18, 36]. On the other hand, whether FAS admits an approximation algorithm with a constant approximation factor still remains a problem to solve.

1.6   Hard to Approximate by Extension

It has been shown, and not only for the FAS but also for all classic NP-Complete problems, those originally proven by Karp [41], that all these have a version that is hard to approximate [69]. These new versions are obtained surprisingly easily, by adding the same simple constraint—by such extension the problems become extremely hard to approximate [69].

For FAS this means the following, for this hard to approximate version of FAS, digraph $G = (V, A)$, $S \subseteq A$, and a feedback arc set of size k, where k is a positive integer—"i.e., a subset $R \subseteq A$ of size k that contains an arc of every directed cycle" [69]. The algorithm for this instance of FAS needs to output feedback arc set of size k with a maximal number of arcs from S [69].

This version of FAS is known as Constrained Max Feedback Arc Set (CM-FAS), and cannot be approximated to within $|A|^\epsilon$ in polynomial time for some $\epsilon > 0$ [69]. This hard to approximate thesis holds for every classic NP-Complete problem, and all these hard to approximate versions are maximization problems [69].

1.7   Feedback Arc Set and Tournament

If we restrict the problem of FAS, that is its input, to digraph tournaments, then we are speaking about a problem known as FAST—Feedback Arc Set on a Tournament. This special case is more tractable, although still NP-Complete and computationally hard [2, 11, 15], and offers some finer ways of acquiring solution.

The restricted FAST allows for a type of algorithm known as polynomial time approximation scheme (PTAS), and when the problem is generalized with arc

Fig. 1.2 Eulerian digraph example

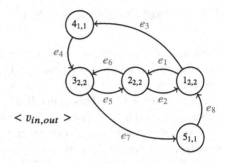

weights, PTAS still holds [44]. When we are speaking about algorithms that are fixed-parameter tractable, FAST does admit it, which is not surprising given it is a special case of FAS for which the same holds (in dense directed graphs which are at most $n^{1+o(1)}$ arcs away from a tournament) [57]. The algorithm is devised for a weighted instance (sometimes also called FAS with probability constraints) of FAST, it is deterministic in nature and sub-exponential in efficiency [42].

The problem of FAS on tournaments can also be stated as a bipartite tournament, this instance of the problem is NP-Complete as well [30], and has been shown to be as such not that long ago, namely in 2007.

1.8 Feedback Arc Set and Eulerian Digraph

The historical problem of Seven Bridges of Königsberg (Kaliningrad) and search for the solution by Leonhard Euler is well known. This work struck the foundation of the graph theory and topology[2] [63]—with Euler referencing Leibniz for his "new kind of geometry without measurement or magnitudes" [45, 59].

Which brings us to a notion of Euler path (sometimes called trail or walk), that represents path which visits each undirected graph edge only once (vertices are traveled at least once) [53, 59]. While Euler graph represents a graph where "every node has even degree" [50]—on the directed graph "the in-degree and the out-degree of each vertex are equal," this is Eulerian digraph [56]. For an example of Eulerian digraph please consult Fig. 1.2—equality of in and out-degree holds, with Eulerian cycle being $\big((1, 2), (2, 1), (1, 4), (4, 3), (3, 2), (2, 3), (3, 5), (5, 1)\big)$.

It has been found that the problem of minimum Feedback Arc Set is on such a graph, namely Eulerian digraph, NP-Hard [56]. Additionally, considering the link between MFAS and the problem called minimum Recurrent Configuration (MRC), NP-Hardness of MRC on general digraphs, as well as on Eulerian digraphs, has been established as well [56].

---

[2] Encyclopedia Britannica: Königsberg bridge problem by Stephan C. Carlson. Accessed on 9th of March 2022 at: https://www.britannica.com/science/Konigsberg-bridge-problem.

1.9   Subset Feedback Arc Set

A variation of the problem exists, a generalized version to be exact, where input is extended with a subset S (of nodes, and arcs), and one needs to find an arc set A' that consists of at least one arc, in $G = (V, A)$, from every directed cycle which intersects with S [20]. This generalized version of FAS is known under a name Subset-FAS, and is NP-Hard even for small subsets (≥ 2) [20].

If subset S consists of a single vertex, then it is possible to solve Subset-FAS in polynomial time by min-cut computation [20]. If one is dealing with undirected graph, then the problem remains NP-Hard even for one element subset S [20].

It has been shown that by computing minimum cuts the weighted Subset-FAS (in an undirected graph) can be approximated by an algorithm which achieves approximation factor of 2—the algorithm has polynomial run-time [21]. "This implies a Δ-approximation algorithm for the subset Feedback Vertex Set problem, where Δ is the maximum degree in G." [21]

1.10   Graph Bipartization

There is a problem where one must remove minimum number of edges/arcs, or vertices, so that the remaining graph is bipartite [22, 29] (bi-graph), that is vertices of a graph can by such removal be divided into two sets R and L, that are disjoint and independent [17]. "A bipartite graph cannot contain an odd cycle, a cycle of odd length." [17]

Such graphs, for example, see [58] and [31], and the problem of Bipartization itself is of interest in terms of Feedback Arc Set problem [22, 29]. One could search for a set with minimum cardinality of arcs that will break all cycles in a collection of cycles C, in a digraph G—the same could be done on a graph (G, w), with w representing non-negative function for arc weights [22]. If this set of cycles C contains all cycles of G that are odd in their length, then we are dealing with the problem of graph Bipartization [22].

1.11   Maximum Acyclic Subgraph—Dual

Dual of the MFAS, known as Maximum Acyclic Subgraph, where one needs to find a maximum set of arcs A' with whose removal graph $G' = (V, A')$ becomes acyclic is APX-complete [32, 55]—and if $|A| = \Theta(|V|^2)$ PTAS that runs in $n^{O(\frac{1}{\epsilon^2})}$ can be achieved [5]. MAS can be approximated to at least

$$\left(\frac{1}{2} + \Omega\left(\frac{1}{\sqrt{\Delta(G)}} \right) \right) |A| \tag{1.2}$$

where $\Delta(G)$ represents maximum vertex degree of G [9]. An improvement of the previous algorithm for an unweighted version of the problem, with a similar bound guarantee and in certain cases improved algorithm complexity, is published in [33]. For instances of MAS where optimal value is $1 - \epsilon$, for a very small ϵ, random algorithm can be beaten with the MFAS algorithm [62] which yields approximation ratio of

$$\frac{1}{2} + \Omega\left(\frac{1}{\log n \log\log n}\right) \tag{1.3}$$

while for instances where optimum is not as close to 1, random guarantee is difficult to beat (as is seen from [9]) [12].

1.12   Feedback Arc and Vertex Set

Procedures have been presented (in [20]) with which one can reduce between Feedback Arc and Vertex Set (FVS) problems. In these reductions, there is a "one-to-one correspondence between feasible solutions and their corresponding costs." [22] This means that approximate solution of the first problem translates to approximate solution of the second one [22].

It has also been shown how to perform reductions between feedback set and feedback subset problems (discussed in Sect. 1.9), and vice versa, while at the same time preserving costs of the feasible solutions [22]. In the continuation we will give reductions for FAS and FVS.

FAS $\preccurlyeq$ FVS [22] Given the graph $G = (V, E)$, construct its directed line-graph $G' = (V', E')$ as follows:

1. Set $V'(G') = E(G)$.
2. An arc in $E'(G')$ connects the vertices $(v_1 \rightarrow v_2) \in V'(G')$ and $(v_3 \rightarrow v_4) \in V'(G')$ iff $v_2 = v_3$.
3. In the weighted version, a correspondence among the weights of the arcs of G and the weights of the "vertices" of the corresponding graph G' is required as follows: the weight of the "vertex" $(v_1 \rightarrow v_2) \in V'(G')$ is equal to the weight of the arc $(v_1 \rightarrow v_2) \in E(G)$.
   A subset of arcs $F \subseteq E(G)$ is a feedback arc set for G iff it is a feedback vertex set for G'.

FVS $\preccurlyeq$ FAS [22] Given the graph $G = (V, E)$, construct a graph $G' = (V', E')$ as follows:

1. For every $v \in V(G)$ insert in $V'(G')$ two vertices v_1 and v_2.
2. For every v_1 and v_2 inserted in $V'(G')$ after splitting a vertex $v \in V(G)$ insert in $E'(G')$ an arc $(v_1 \rightarrow v_2)$, all the arcs that enter v in G connecting them to v_1 in G', and all the arcs that emanate from v in G emanating them from v_2 in G'.

3. In the case of weights we also have: for every arc $e' = [(v_1 \rightarrow v_2) \mid v \in V(G)]$, set $w(e') = w(v)$. All other arcs in $E'(G')$ have infinite weight.

   This establishes a one-to-one correspondence between the finite weighted feedback arc sets of the new graph G' and the feedback vertex sets of the original graph G.

Reductions presented above, both FAS $\preccurlyeq$ FVS and FVS $\preccurlyeq$ FAS, can be performed in linear time—"and therefore these problems can be regarded as different representations of the same problem" [20, 22].

1.13 The Forcing Problem

There is a connection (shown in [54]) between Feedback Arc Set and the Forcing Problem (FP) in square grids. If one has a graph G which is admitting M (represents a perfect matching), the forcing number of such perfect matching is "the smallest number of arcs in a subset $S \subset M$"—where S is not in any other perfect matching [22]. "A subset S having this property is said to force M" [22].

If one has a bi-graph G, and its perfect matching M, where "the maximum size of a collection of edge disjoint cycles equals the minimum size of a feedback set" [54] (called the cycle-packing property—if the graph is undirected, then the cycle-packing property holds "if every orientation of the edges results in a directed graph with the cycle-packing property" [54]), then a digraph $D(M)$ can be constructed with the same set of vertices that G has and those vertices can be partitioned into two sets (A and B) [22, 54]. If $e \in M$, then direct e from set A to set B, and if $e \notin M$, then direct e oppositely [54].

It is clear that in such a situation there is a "one-to-one correspondence between alternating cycles in M and directed cycles in $D(M)$," with two-cycles in $D(M)$ intersecting on an edge or not intersecting at all [54]. "There is also a natural correspondence between forcing sets in M and feedback sets in $D(M)$." [54] For every feedback set that is in $D(M)$ there exists a corresponding forcing set in M of equal cardinality—the opposite also holds true (lemma 3 from [54]) [54].

1.14 Practical Relevance

Considering a standing of FAS in computing, and consequently MFAS, the significance in the theory of computation is clear. It is one of those hard to solve problems that by its characteristics bears significance not only for its "brothers in arms" problems but also for the computing at large. On the other hand, is this practically relevant? These hard to solve problems, their complexity classes, and everything surrounding these topics seems only academic in nature, but it actually is not so.

These intractable problems, more often than one would think, do find their way into practical situations, and when that happens, then we are quite interested in efficient and optimal way of finding a solution. The same is true for the problem of FAS, it is simplistic in nature, hard to solve, and actually quite often found in practice.

Situations where the problem of FAS emerges are given in the following text. Two groups are given, in the group of "comp" there is a list of FAS applications were computing and electronics are of a main concern, while in the group of "othr" there is a list of applications where computing and electronics need not be a primary focus, but where computers might be the means of delivering the result.

COMP   **Machine** learning [10, 42, 52], **Search** engine ranking [10, 25, 42, 52], **Layered** graph drawing [16, 44, 51], **Prog**ram verification [23, 60], **Test**ing electronic circuits [20, 35, 46], **Deadlock** resolution [20, 23, 46, 47], **Misinf**ormation removal and label propagation in social networks [46, 64], **Re-timing** synchronous circuitry [35], **Comp**utational biology and neuroscience [8, 35], **Net**work analysis [35], **Data**base systems [8], Very large scale integration (**VLSI**) chip design [8, 60], **Control** structure of computer programs [24], **Code** optimization [24], **Crypto**graphy [60], Data flow analysis (**DFA**) [19, 24, 38, 65], Circuit design (**CD**) [47].

OTHR   **Voting** [10, 24, 33, 42, 52], **Tour**naments [11, 25, 42, 44], **Find**ing the most consistent rankings in psychology [25, 42, 46], **Inf**ormation system subsystems sequential ordering [46, 66], **Tear**ing in chemical engineering [7], **Map**ping between ontologies [52], Determining hierarchy of sectors of an economy (**DHSE**) [33], **Ancestry** relationships [3, 33], **Analysis** of systems with feedback [3, 16, 33, 66], **Scheduling** [16, 33, 44, 47], **Constraint** satisfaction [16, 23], Group **rank**ing and statistics [24, 25, 42–44], Decision making (**DM**) [24], **Circadian** rhythm in animals and plants [37], **Apoptosis** of cells and cancer growth in human tissues [1], **Bayesi**an inference [23, 44], Network integrity (**NI**) [68], **Fault** tolerance [47].

There are many practical applications of FAS. Some are more practical than others, ranging from very important in practice to somewhat important, and some are more similar to each other than others. We could perhaps argue that some items in the above classification could change the group, depending on the corner from which we look at these, it is, however, evident that FAS has wide application base both in computing, which is not surprising considering the ubiquity of technology in our lives, and in other disciplines, which is again logical considering the breadth of interconnectedness and self-reference.

Therefore having an efficient, and solution-wise precise, algorithm for the problem of FAS is important. Still, considering pervasiveness of self-reference in the world we live in, the above list of FAS practical relevance is probably not exhaustive.

References

1. Aguda, B.D., Kim, Y., Piper-Hunter, M.G., Friedman, A., Marsh, C.B.: Microrna regulation of a cancer network: consequences of the feedback loops involving mir-17-92, e2f, and myc. Proc. Nat. Acad. Sci. U.S.A. **105**, 19678–19683 (2008)
2. Alon, N.: Ranking tournaments. SIAM J. Discrete Math. **20**(1), 137–142 (2006)
3. Aneja, Y.P., Sokkalingam, P.T.: The minimal feedback arc set problems. Inform. Syst. Oper. Res. **42**(2), 107–112 (2004)
4. Ariyoshi, H., Higashiyama, Y.: A heuristic algorithm for the minimum feedback arc set problem. Res. Inst. Math. Anal. **427**, 112–130 (1981). Kyoto University Research Information Repository (Departmental Bulletin Paper)
5. Arora, S., Frieze, A., Kaplan, H.: A new rounding procedure for the assignment problem with applications to dense graph arrangement problems. Math. Program. **92**(1), 1–36 (2002)
6. Ausiello, G., D'Atri, A., Protasi, M.: Structure preserving reductions among convex optimization problems. J. Comput. Syst. Sci. **21**(1), 136–153 (1980)
7. Baharev, A., Schichl, H., Neumaier, A., Achterberg, T.: An exact method for the minimum feedback arc set problem. ACM J. Exp. Algorithm. **26**, 1–28 (2021)
8. Bang-Jensen, J., Maddaloni, A., Saurabh, S.: Algorithms and kernels for feedback set problems in generalizations of tournaments. Algorithmica **76**(2), 320–343 (2015)
9. Berger, B., Shor, P.W.: Approximation algorithms for the maximum acyclic subgraph problem. In: SODA '90: Proceedings of the First Annual ACM-SIAM Symposium on Discrete Algorithms, pp. 236–243. Society for Industrial and Applied Mathematics (1990)
10. Bessy, S., Bougeret, M., Krithika, R., Sahu, A., Saurabh, S., Thiebaut, J., Zehavi, M.: Packing arc-disjoint cycles in tournaments. Algorithmica **83**, 1393–1420 (2021)
11. Charbit, P., Thomassé, S., Yeo, A.: The minimum feedback arc set problem is NP-hard for tournaments. Comb. Probab. Comput. **16**(1), 1–4 (2006)
12. Charikar, M., Makarychev, K., Maka, Y.: On the advantage over random for maximum acyclic subgraph. In: 48th Annual IEEE Symposium on Foundations of Computer Science (FOCS'07), pp. 625–633. IEEE, Piscataway (2007)
13. Charon, I., Guénoche, A., Hudry, O., Woirgard, F.: New results on the computation of median orders. Discrete Math. **165–166**, 139–153 (1997)
14. Charon, I., Hudry, O.: A branch-and-bound algorithm to solve the linear ordering problem for weighted tournaments. Discrete Appl. Math. **154**(15), 2097–2116 (2006)
15. Conitzer, V.: Computing slater rankings using similarities among candidates. In: AAAI'06: Proceedings of the 21st National Conference on Artificial Intelligence, vol. 1, pp. 613–619. AAAI Press (2006)
16. Demetrescu, C., Finocchi, I.: Combinatorial algorithms for feedback problems in directed graphs. Inform. Proc. Lett. **86**(3), 129–136 (2003)
17. Diestel, R.: Graph Theory, Graduate Texts in Mathematics, vol. 173, 3rd edn. Springer, Heidelberg (2005)
18. Dinur, I., Safra, S.: On the hardness of approximating vertex cover. Ann. Math. **162**(1), 439–485 (2005)
19. Domínguez-García, V., Pigolotti, S., Muñoz, M.A.: Inherent directionality explains the lack of feedback loops in empirical networks. Sci. Rep. **4**(1) (2014)
20. Even, G., Naor, J.S., Schieber, B., Sudan, M.: Approximating minimum feedback sets and multicuts in directed graphs. Algorithmica **20**(2), 151–174 (1998)
21. Even, G., Naor, J.S., Schieber, B., Zosin, L.: Approximating minimum subset feedback sets in undirected graphs with applications. SIAM J. Discrete Math. **13**(2), 255–267 (2000)
22. Festa, P., Pardalos, P.M., Resende, M.G.C.: Feedback set problems. In: Handbook of Combinatorial Optimization, vol. A, pp. 209–258. Springer (1999)
23. Festa, P., Pardalos, P.M., Resende, M.G.C.: Algorithm 815: Fortran subroutines for computing approximate solutions of feedback set problems using grasp. ACM Trans. Math. Softw. **27**(4), 456–464 (2001)

24. Flood, M.M.: Exact and heuristic algorithms for the weighted feedback arc set problem: a special case of the skew-symmetric quadratic assignment problem. Networks **20**(1), 1–23 (1990)
25. Fomin, F.V., Lokshtanov, D., Raman, V., Saurabh, S.: Fast local search algorithm for weighted feedback arc set in tournaments. In: AAAI'10: Proceedings of the Twenty-Fourth AAAI Conference on Artificial Intelligence, vol. 24, pp. 65–70. AAAI Press (2010)
26. Garey, M.R., Johnson, D.S.: Computers and Intractability: A Guide to the Theory of NP-Completeness. W. H. Freeman, San Francisco (1979)
27. Gavril, F.: Some NP-complete problems on graphs. In: 11th Conference on Information Sciences and Systems, pp. 91–95. Johns Hopkins University (1977)
28. Grotschel, M., Junger, M., Reinelt, G.: Acyclic subdigraphs and linear orderings: polytopes, facets, and a cutting plane algorithm. In: Graphs and Order. NATO ASI Series (Series C: Mathematical and Physical Sciences), vol. 147, pp. 217–264. Springer, Dordrecht (1985)
29. Guo, J., Gramm, J., Huffner, F., Niedermeier, R., Wernicke, S.: Compression-based fixed-parameter algorithms for feedback vertex set and edge bipartization. J. Comput. Syst. Sci. **72**(8), 1386–1396 (2006)
30. Guo, J., Huffner, F., Moser, H.: Feedback arc set in bipartite tournaments is NP-complete. Inform. Proc. Lett. **102**(2–3), 62–65 (2007)
31. Gupta, S.: Feedback arc set problem in bipartite tournaments. Inform. Proc. Lett. **105**(4), 150–154 (2008)
32. Guruswami, V., Manokaran, R., Raghavendra, P.: Beating the random ordering is hard: inapproximability of maximum acyclic subgraph. In: 2008 49th Annual IEEE Symposium on Foundations of Computer Science, pp. 573–582. IEEE, Piscataway (2008)
33. Hassin, R., Rubinstein, S.: Approximations for the maximum acyclic subgraph problem. Inform. Proc. Lett. **51**(3), 133–140 (1994)
34. Haynes, T.W., Hedetniemia, S., Slater, P.: Fundamentals of Domination in Graphs, 1st edn. CRC Press, Boca Raton (1998)
35. Hecht, M.: Exact Localisations of Feedback Sets. Theory Comput. Syst. **62**(5), 1048–1084 (2017)
36. Hecht, M., Gonciarz, K., Horvát, S.: Tight localizations of feedback sets. ACM J. Exp. Algorithmics **26**, 1–19 (2021)
37. Ingalls, B.P.: Mathematical Modeling in Systems Bology: An Introduction. MIT Press, Cambridge, MA (2013)
38. Ispolatov, I., Maslov, S.: Detection of the dominant direction of information flow and feedback links in densely interconnected regulatory networks. BMC Bioinform. **9**(424) (2008)
39. Kamae, T.: Notes on a minimum feedback arc set. IEEE Trans. Circuit Theory **14**(1), 78–79 (1967)
40. Kann, V.: On the approximability of np-complete optimization problems. Ph.D. Thesis, Royal Institute of Technology, Stockholm, Sweden (1992)
41. Karp, R.M.: Reducibility among combinatorial problems. In: Complexity of Computer Computations, IRSS, pp. 85–103. Springer (1972)
42. Karpinski, M., Schudy, W.: Faster algorithms for feedback arc set tournament, Kemeny rank aggregation and betweenness tournament. In: Algorithms and Computation. Lecture Notes in Computer Science, vol. 6506, pp. 3–14. Springer, Berlin (2010)
43. Kemeny, J.G.: Mathematics without numbers. Daedalus **88**(4), 577–591 (1959)
44. Kenyon-Mathieu, C., Schudy, W.: How to rank with few errors. In: Proceedings of the Thirty-Ninth Annual ACM Symposium on Theory of Computing—STOC '07, pp. 95–103. ACM Press (2007)
45. Kruja, E., Marks, J., Blair, A., Waters, R.: A short note on the history of graph drawing. In: Graph Drawing. Lecture Notes in Computer Science, vol. 2265, pp. 272–286. Springer, Berlin (2002)
46. Kudelić, R., Ivković, N.: Ant inspired monte carlo algorithm for minimum feedback arc set. Expert Syst. Appl. **122**, 108–117 (2019)

47. Kuo, C.J., Hsu, C.C., Lin, H.R., Chen, D.R.: Minimum feedback arc sets in rotator and incomplete rotator graphs. Int. J. Found. Comput. Sci. **23**(04), 931–940 (2012)
48. Lawler, E.: A comment on minimum feedback arc sets. IEEE Trans. Circuit Theory **11**(2), 296–297 (1964)
49. Lucchesi, C.L., Younger, D.H.: A minimax theorem for directed graphs. J. Lond. Math. Soc. **s2-17**(3), 369–374 (1978)
50. Mallows, C.L., Sloane, N.J.A.: Two-graphs, switching classes and euler graphs are equal in number. SIAM J. Appl. Math. **28**(4), 876–880 (1975)
51. Marik, R.: On multitree-like graph layering. In: Studies in Computational Intelligence. Studies in Computational Intelligence, vol. 689, pp. 595–606. Springer, Berlin (2017)
52. Misra, P., Raman, V., Ramanujan, M.S., Saurabh, S.: A polynomial kernel for feedback arc set on bipartite tournaments. Theory Comput. Syst. **53**(4), 609–620 (2013)
53. Mohan, S.: Eulerian Path and Tour Problems. Wiley Encyclopedia of Operations Research and Management Science (2011)
54. Pachter, L., Kim, P.: Forcing matchings on square grids. Discrete Math. **190**(1–3), 287–294 (1998)
55. Papadimitriou, C.H., Yannakakis, M.: Optimization, approximation, and complexity classes. J. Comput. Syst. Sci. **43**(3), 425–440 (1991)
56. Perrot, K., Pham, T.V.: Feedback arc set problem and NP-hardness of minimum recurrent configuration problem of chip-firing game on directed graphs. Ann. Comb. **19**(2), 373–396 (2015)
57. Raman, V., Saurabh, S.: Parameterized algorithms for feedback set problems and their duals in tournaments. Theor. Comput. Sci. **351**(3), 446–458 (2006)
58. Raman, V., Saurabh, S., Sikdar, S.: Improved exact exponential algorithms for vertex bipartization and other problems. In: ICTCS: Italian Conference on Theoretical Computer Science. LNCS, vol. 3701, pp. 375–389. Springer, Berlin (2005)
59. Sachs, H., Stiebitz, M., Wilson, R.J.: An historical note: Euler's Konigsberg letters. J. Graph Theory **12**(1), 133–139 (1988)
60. Schwikowski, B., Speckenmeyer, E.: On enumerating all minimal solutions of feedback problems. Discrete Appl. Math. **117**(1-3), 253–265 (2002)
61. Seshu, S., Reed, M.B.: Linear Graphs and Electrical Networks. Addison-Wesley, Reading, MA (1961)
62. Seymour, P.D.: Packing directed circuits fractionally. Combinatorica **15**(2), 281–288 (1995)
63. Shields, R.: Cultural topology: the seven bridges of Konigsburg, 1736. Theory Culture Soc. **29**(4-5), 43–57 (2012)
64. Simpson, M., Srinivasan, V., Thomo, A.: Efficient computation of feedback arc set at web-scale. Proc. VLDB Endowment **10**(3), 133–144 (2016)
65. Xu, Y.Z., Zhou, H.J.: Optimal segmentation of directed graph and the minimum number of feedback arcs. J. Stat. Phys. **169**(1), 187–202 (2017)
66. Younger, D.: Minimum feedback arc sets for a directed graph. IEEE Trans. Circuit Theory **10**(2), 238–245 (1963)
67. Zhao, J.H., Habibulla, Y., Zhou, H.J.: Statistical mechanics of the minimum dominating set problem. J. Stat. Phys. **159**(5), 1154–1174 (2015)
68. Zhao, J.H., Zhou, H.J.: Optimal Disruption of Complex Networks (2016)
69. Zuckerman, D.: Np-complete problems have a version that's hard to approximate. In: [1993] Proceedings of the Eigth Annual Structure in Complexity Theory Conference, pp. 305–312. IEEE Comput. Soc. Press (1993)

Part II
Feedback Arc Set and Algorithms Thereof

Chapter 2
Introductory Remarks

2.1 Methodology

In this book part we will make a compendium of algorithms primarily for FAS, and MFAS. There are some variations of the problem, and there are also some interesting borderline problems, and algorithms, that will also be to a certain extent covered, depending on their significance.

However, it is not our intention to make this chronological review too wide, since this would consequently end up in a compendium of at least all NP-Complete problems, and their subsequent algorithms. There is always a possibility of missing an algorithm here, and there, and for this we upfront apologize to authors of such papers—"time" will be a correction of that, as typically is in science.

For every paper, and for every algorithm, we will present a briefest summary possible of the most important knowledge from the paper being discussed. In this way the reader will be able to get a gist of the algorithm at a glance, and will be able to quickly make analysis, and comparisons of their own, without having to deal with the research which he is not currently conducting—for everything else there is a thorough reading of each paper, which should be easy to find, since extra care was devoted to making literature data precise.[1]

In this way the reader can easily, and very quickly, jump through papers, download them, and read them. If the pseudo-code for the algorithm is not available, so as not to repeat the text of the procedure in question, the reader is directed to the paper itself which will typically be easy to find.

---

[1] Online supplementary material, mentioned in the preface, has almost entire monograph literature digitally referenced.

© The Author(s), under exclusive license to Springer Nature Switzerland AG 2022
R. Kudelić, *Feedback Arc Set*, SpringerBriefs in Computer Science,
https://doi.org/10.1007/978-3-031-10515-9_2

2.2   Cited Literature

Considering that in the following sub-chapters, of the next chapter of the book, that deal with particular algorithms, each sub-chapter will typically have information acquired from one source only, we will not use any citation rules, aside from giving a citation in a section title.

The purpose of this being, avoiding clutter of the chapter text, making information flow seamless, and making the text as elegant and easy to follow, and as understandable as possible—since it is quite clear, and without a doubt, from where the information and text was taken. The same rule will be applied to figures, tables, and of course algorithms as well, sub-chapter title has the citation.

For situations where given information goes outside the main article, that is article of the sub-chapter (paper in question that is being summarized), usual citation will be given. In this way we will provide literature referencing on the one side, and text focused on the content on the other.

2.3   About the History

One should take note that the historical review presented in this book does not necessarily mention all problems, issues, algorithms for other worthy problems, presented in the papers we are conducting a review of.

The main focus is on Feedback Arc Set and everything directly related to that problem. If it is of consequence for FAS, then it will be mentioned, or we will direct the reader to the paper itself for additional details.

As an exception, here and there, some strongly related problems have been covered, this gives the review additional relevance by discussing a number of problems that are on the outskirts of FAS, but relevant to FAS. This complements the material about FAS by giving a more complete picture, and also directs the reader to the objects on the horizon, to the problems for further study.

Considering the constraint of the number of pages of this book, a curated selection of papers, with algorithms thereof, is given. Such a selection will give both the breadth of traveling through time and depth of research at certain points in time.[2]

If an issue is at some distance from FAS, then the content presented in the book will not be untidied with such information, and we would kindly direct the reader to do a search for a problem he is looking for.

---

[2] On the web-page directed to in the preface the reader can find additional material not covered in the book, including references to all papers that could be of interest but were not covered.

Chapter 3
Papers and Algorithms

3.1 On Directed Graphs and Integer Programs[1] [128] 1960

This is the first paper that deals strongly with FAS. Previous papers were dealing with the fledgling problem of FAS, and not as noteworthy.

3.2 Minimum Feedback Arc Sets for a Directed Graph [133]—June 1963

This paper deals specifically with FAS, that is with its minimization version. Until this paper there were two names suggested for MFAS: minimal feedback cut-set, and minimal chord set. Considering shortcomings of both, and the fact that the name was not yet standardized, Younger suggested a name that has been accepted by scientific community, and has stood the test of time, namely Minimum Feedback Arc Set.

The paper establishes a relationship between feedback arcs, and order, including admissible ordering with feedback sets that may be less than optimal. From determined relationship Younger has developed some simple characteristics of sets that would qualify, and has developed properties of sequential orderings with which minimum sets are determined.

Younger has in his paper also described the algorithm for finding minimum feedback sets. Series of steps for such an algorithm are as follows:

---

[1] It was not possible to acquire it and include it in this book. For more details on the issue please consult supplementary online material, mentioned in the preface.

© The Author(s), under exclusive license to Springer Nature Switzerland AG 2022
R. Kudelić, *Feedback Arc Set*, SpringerBriefs in Computer Science,
https://doi.org/10.1007/978-3-031-10515-9_3

1. **Preliminary**: Obtain an admissible ordering R, relabeling the graph according to this ordering. Find the feedback arc set determined by R and note the number of elements in this set $Q(R)$.

2. **Branching operation**: For each feedback arc, perform a single node perturbation on R to establish the down sequent corresponding to this arc. Unite the nodes of this down sequent into a single node, eliminating loops of two arcs created by this union. Add the number of such loops to an index number I, where I indicates for a given perturbed ordering the total number of such cancellations.

3. **Evaluation and transfer**: (a) If $Q(R) - I < 0$, reject the down sequent under consideration. Return to the next feedback arc in step 2, resetting I to the number it was before consideration of this rejected down sequent. If all feedback arcs at a given branching point have been examined, then proceed back to the next arc of the branching point of the next higher level; if at the first level, proceed to step 4. (b) If $Q(R) - I \geq 0$, rearrange the perturbed ordering until it is admissible. If the number of elements in the feedback arc set determined by the perturbed ordering $Q(R_{pert}) < Q(R) - I$, consider it the new reference ordering, and begin the program again at step 1. If $Q(R_{pert}) = Q(R) - I$, list R_{pert} on a list of potentially optimum F representatives. If $Q(R_{pert}) \geq Q(R) - I$, note the feedback arcs determined by this ordering and for this new level of branching, proceed back to step 2.

4. **Final processing**: From the reference ordering and all others on the potentially optimum list, find all optimum F representatives by the following equivalences. "Given an optimum ordering R for a directed graph G, let G_1 and G_2 be consecutive sub-graphs of n_1 and n_2 nodes such that the highest numbered node in G_1 is one less than the lowest numbered node in G_2. Then, (a) $c_{G_1 G_2} \geq c_{G_2 G_1}$, (b) if $c_{G_1 G_2} = c_{G_2 G_1}$, then the ordering R' is also optimum, where for each node i of G

$$R'(i) = R(i), \qquad i \notin G_1 \text{ or } G_2$$
$$R'(i) = R(i) - n_1, \quad i \in G_2 \qquad\qquad (3.1)$$
$$R'(i) = R(i) + n_2, \quad i \in G_1$$

The condition $c_{G_1 G_2} \geq c_{G_2 G_1}$ is necessary for R to be optimum, this condition must also hold for all other possible G_1, G_2 as well"—for each optimum ordering, enumerate the corresponding feedback arc set. Forward and backward arcs are, respectively, represented by $c_{G_1 G_2}$, and $c_{G_2 G_1}$. For more details the reader should consult theorem 6, and corresponding figures 2, 3, and 4, of the paper in question.

Efficiency of the algorithm was, as Younger has expressed it, not easy to estimate—the author found, in a few hours by hand calculation, the minimum feedback arc sets for 12-node, and 41-arc graph. For which the arcs were chosen randomly. For a graph too large for a complete search, a partial search can be made by being more selective about the down sequents to be investigated, and for still larger graphs, one may be content with an admissible ordering.

Generally speaking, when graph is large, resource spent is only limitation, as is the case for difficult computational problems. By method of Algorithm 3.2.1, as expounded by [117], any ordering can be converted into an admissible one.

Algorithm 3.2.1 Iterative improvement method for obtaining an admissible ordering

---

let $v_1, \ldots, v_n$ be an initial ordering
improve := *true*
while *improve* = *true* **do**
  improve := *false*
  for $1 \leq i \leq j < k \leq n$ **do**
    $G_1 := G[v_i, \ldots, v_j]$
    $G_2 := G[v_{j+1}, \ldots, v_k]$
    if $N(G_1, G_2) < N(G_2, G_1)$ **then**
      change order to $v_1, \ldots, v_{i-1}, v_{j+1}, \ldots, v_k, v_i, \ldots, v_j, v_{k+1}, \ldots, v_n$
      renumber vertices according to the new order
      improve := *true*
    end if
  end for
  $F :=$ set of leftward arcs with respect to current ordering
  for all $e \in L$ **do**
    if $H(V, (E \setminus F) \cup \{e\})$ is acyclic **then**
      use topological sorting to reorder H {consult [35]}
      renumber vertices according to new order
      remove from F all arcs that become rightward
      improve := *true*
    end if
  end for
end while
return $v_1, \ldots, v_n$

3.3 Minimum Feedback Arc and Vertex Sets of a Directed Graph [93]—December 1966

This paper deals both with Minimum Feedback Arc Set, and with Minimum Feedback Vertex Set (MFVS). The paper presents a closed form solution for both MFAS, and MFVS problems. Determination of minimum sets for either MFAS or MFVS problem involves the expansion of an n-th order permanent, where n is the number of the graph vertices, and some algebraic manipulations of the resultant expression subject to the absorption laws of Boolean algebra—generally a graph may possess several minimum sets of either kind.

The algorithm presented in the paper outputs all possible solutions simultaneously. The authors first give an algorithm for graphs without self-loops, and without multiple arcs (multi-graph). Then, the results obtained are generalized for arbitrary graph.

Since the algorithms were not given in a programming friendly manner, without repeating large parts of the paper, intricacies of the algorithms given would be lost, therefore we direct the reader to the paper itself for the exact algorithms the authors have presented in order to tackle MFAS.

3.4 A Heuristic Algorithm for the Minimum Feedback Arc Set Problem [8]—June 1981

The authors have proposed an approximate algorithm for the identification of near optimal feedback arc sets of a directed graph. By exploiting Depth-First Search (DFS) characteristics, they have constructed an algorithm for MFAS. It has been shown that introduction of a modified DFS with vertex-selection-rules, and interchange-rules, for graph links is significant in improving the size of a feedback arc set.

Approach for finding MFAS presented in this paper is mainly based upon well known work of Robert Tarjan that can be found in [126]—Tarjan deals with arguing a value of backtracking as a technique for solving various problems [126].

Pseudo-code for modified DFS that is a part of a heuristic algorithm for the Minimum Feedback Arc Set can be seen in Algorithm 3.4.1. This procedure has algorithm time complexity of $O(n + m + m^2)$, and a space complexity of $O(n \cdot m)$. A minimization technique which calls Algorithm 3.4.1 can be seen in Algorithm 3.4.2—its time complexity is $O(n^2 m^2)$. Since minimality of minimization is not guaranteed, in order to rectify this, minimization technique can be followed by a call to a minimality check routine—see Algorithm 3.4.3 for a DFS based check routine. This minimality check procedure has a time complexity of $O(n^2 m^2)$, and a space complexity of $O(n \cdot m)$.

Selection Rules A vertex v_j to be visited during a DFS is selected under the following rules: 1) the vertex v_j has the minimum in-degree $d_0^-(j)$ among the adjacent vertices v_{jk}'s s.t. $(v_i, v_{jk}) \in E$, and 2) if there are two or more candidates, v_j has the maximum out-degree $d_0^+(j)$.

Algorithms 3.4.1, 3.4.2, and 3.4.3 are heuristics, and therefore are not guaranteeing solution quality. Nevertheless, it is possible to continuously improve and approximate the solution by recursively calling procedure Algorithm 3.4.3. Such an algorithm have the authors of the paper presented in Algorithm 3.4.4. This procedure works by deletion of feedback arcs, one by one, and thus continually refining a solution, and checking for its minimality. Procedure for MFAS approximation has an estimated time complexity of $O(n^2 m^2)$, and space complexity of $O(n \cdot m)$.

Algorithm 3.4.1 Modified depth first search—**procedure** MDFS($G = (V, E)$)

---

integer array $N(|V|), M(|V|)$
arc set $\Omega_T, \Omega_B, \Omega_{CF}, E_B$
integer value $C1, C2$

procedure $BACKTRACK(vertex\ set\ V_0,\ vertex\ v_i,\ vertex\ v_k)$
if there exists a vertex $v_i \in V_0$ for which the selecting rule holds **then**
   $M(v_j) := C1 := C1 + 1$
   $V_0 := V_0 - \{v_j\}$
   $\Omega_T := \Omega_T \cup \{< v_i, v_j >\}$
   $BACKTRACK(V_0, v_j, v_i)$
   $E_B(v_i) := E_B(v_i) \cup E_B(v_j)$
   $N(v_j) := C2 := C2 + 1$
else
   for all vertex $v_j \notin V_0, < v_i, v_j >\in E$ **do**
     if $N(v_j) = 0$ **then**
        $\Omega_B := \Omega_B \cup \{< v_i, v_j >\}$
        $E_B(v_i) := E_B(v_i) \cup \{< v_i, v_j >\}$
     else if $< v_i, v_j >\notin \Omega_T$ **then**
        $\Omega_{CF} := \Omega_{CF} \cup \{< v_i, v_j >\}$
        $E_B(v_i) := E_B(v_i) \cup E_B(v_j)$
     end if
   end for
end if
end $BACKTRACK$

$V_0 := V$
while there exists a vertex $v_i \in V_0$ for which the selecting rule holds **do**
   $M(v_i) := C1 := C1 + 1$
   $V_0 := V_0 - \{v_i\}$
   $BACKTRACK(V_0, v_i, Dummy)$
end while

Procedure, that is the entire algorithm, has been implemented on a FACOM M-200. Empirical testing has been conducted on an input graph called Directed Star Polygon (DSP—definition 5 of the paper in question). The algorithm was, for every particular DSP, run multiple times, so as to better evaluate efficiency. Experimental data shows that procedure Algorithm 3.4.4 generates close to optimal solutions within deciseconds—whether this would hold for any graph, the authors do not give a hard guarantee. For finding tearing and matrix associated MFAS and MFVS of a directed graph one can consult [18, 29, 63].

Algorithm 3.4.2 Minimization technique—**procedure** LEASTFAS($G = (V, E)$)

arc set $MFAS$, Ω_B
integer array $k(|V|)$

procedure $REFTREE(G' = (V, E'))$
$MDFS(G')$
if $\Omega_B \neq \emptyset$ **then**
    compute $k(e)$ for each $e \in \Omega_T$ and identify RT-edge e_r
end if
if $k(e_r) > 0$ **then**
    $MFAS := MFAS \cup F_{CF,eff}(e_r) \cup \{e_r\}$
else
    $MFAS := MFAS \cup F_B(e_r)$
    $E' := E' - MFAS$
    $REFTREE(G')$
end if
end $REFTREE$

$MFAS := \emptyset$
$REFTREE(G)$

3.5 A Branch and Bound Algorithm for the Acyclic Subgraph Problem [78]—December 1981

The paper deals with the Acyclic Subgraph Problem (ASP), as the author called it, which is the dual problem of the Feedback Arc Set. A depth-first search (DFS) type of algorithm has been developed. The algorithm works by lexicographically enumerating all permutations, and then avoiding segments that are known by easy tests not to have optimal completion.

DFS was used since it was more convenient, when recursion is of importance, and it used less memory, which was at the time of much more importance than in today's world. To determine lower bound the authors have used, and developed their own algorithm for finding a large collection of disjoint cycles. This algorithm works in $O(n^4)$ time space, and in $O(n^3)$ memory space.

This DFS type of algorithm can adequately solve a problem of sizes ranging from 25 to 34 nodes, such variability comes from the nature of the problem being tackled. During experimental stage of the research it has been determined (by testing on data representing Netherlands economy, aggregations of random rankings, and randomized problems with Bernoulli, uniform, and normal distributions) that the algorithm performance ranges from $1.7s$ to $20s$—which was determined with PASCAL program on a CDC-Cyber 750 mainframe-class supercomputer.

Global lower bound, for this DFS type of algorithm, of the solution with respect to optimum was always within at least 94%, with the best output of 99.5% for the input-output matrices of the Netherlands economy—showing that real-world problems were in this instance easier to solve. While percentage of total weight on

---

Algorithm 3.4.3 Minimality check—**procedure** APFAS#1$(G = (V, E))$

integer array $N(|V|), M(|V|)$
arc set $MFAS, MFAS'$

procedure $DFS(G'' = (V, E''), vertex\ v_i, vertex\ v_k)$
for all vertex v_j, $< v_i, v_j >\in E''$ **do**
  if $M(v_j) = 0$ **then**
    $M(v_j) := 1$
    $DFS(G'', v_j, v_i)$
    $N(v_j) := 1$
  else if $N(v_j) = 0$ **then**
    $M(\cdot) := 0$
    $N(\cdot) := 0$
    goto cfound {at this point a cycle has been found}
  end if
end for
end DFS

$LEASTFAS(G)$
$MFAS' := MFAS$
$MFAS := \emptyset$
delete every edge $e \in MFAS'$ from a given graph G and let the resultant graph be G'

cfound:
while $MFAS' \neq \emptyset$ **do**
  select and delete any edge e_m from $MFAS'$
  $MFAS := MFAS \cup \{e_m\}$
  $E'' := E' \cup \{e_m\}$
  while any vertex v_i, $M(v_i) = 0$ **do**
    $M(v_i) := 1$
    $DFS(G'', v_i, Dummy)$
  end while
  $MFAS := MFAS - \{e_m\}$
  $E' := E' \cup \{e_m\}$
end while

---

suboptimal arcs was at most 23%, with minimum being 3.1%, and median being 20%. For other comparable algorithms one can look at [38, 55].

3.6 Acyclic Subdigraphs and Linear Orderings: Polytopes, Facets, and a Cutting Plane Algorithm [61]—?? 1985

In this paper, perhaps under a somewhat strange title, the authors are dealing with Acyclic Subdigraph Problem, and the Linear Ordering Problem (LOP). As we know, ASP is a dual version of the primary problem of FAS. The problem is being looked at from a polyhedral point of view, where partial knowledge of the facet structure of

---

Algorithm 3.4.4 MFAS approximation—**procedure** APFAS#2($G = (V, E)$)

---

arc set $MFAS, MFAS'$
integer value K

procedure $REFINE(G = (V, E))$
$MFAS' := MFAS$
while $MFAS' \neq \emptyset$ **do**
    select and delete any edge e from $MFAS'$, delete e from a given graph G and let the resultant
    graph be G'
    $APFAS\#1(G')$
    if $K > |MFAS| + 1$ **then**
        $K := |MFAS| + 1$
        $REFINE(G)$
    end if
end while
end $REFINE$

$APFAS\#1(G)$
$K := |MFAS|$
$REFINE(G)$

---

polytopes leads to both formulation, and implementation of cutting plane algorithm for Acyclic Subdigraph Problem, which is often stated as LOP.

Vertices of acyclic subdigraph polytopes $P_{AC}(D)$ being studied, associated with digraphs $D = (V, A)$, correspond bijectively to the acyclic arc sets $B \subseteq A$. By studying facial structure of these polytopes the authors have derived several facet classes of these polytopes through which, and by combining with heuristic and Branch and Bound techniques, cutting plane algorithm is defined. Graphs being considered are simple, i.e., without loops or parallel edges, and main consideration are digraphs.

The key point of the cutting plane algorithm for Linear Ordering Problem, is to successively solve a number of stronger relaxations of the linear program. A very general pseudo-code of the algorithm is given in Algorithm 3.6.1. Time complexity of the method is $O(n^6)$, although by adjacency list data structure implementation, and with not too many edges, a more efficient practical run-time is expected—linear programs were being solved by MPSX/370.

The algorithm was applied to triangulation problem for input-output tables of the economy of certain countries. Experiments were being run on a SIEMENS 7.865, and computational results were as follows. The algorithm needed approximately 1.5 min for the 44-sector tables, 2.5 min for the 50-sector table, 5 min for the 56-sector tables, and 10 min for the 60-sector tables. This showed that the algorithm favorably compared to alternate approaches of the time, and that a linear programming based method was a sound way of trying to solve the problem in question.

Further investigation revealed that the algorithm could be used to successfully solve even larger instance, of size 80 and perhaps larger, for those instances that resemble real-world problems. But when a problem would be randomly generated,

---

Algorithm 3.6.1 Cutting plane pseudo-code for LOP

$P := \{x \in \mathbb{R}^{n(n-1)} | x_{ij} + x_{ji} = 1, \text{ for all } 1 \leq i < j \leq n, x_{ij} \geq 0, \text{ for all } 1 \leq i, j \leq n\}$
$found := true$

while $found$ **do**
$\quad$ solve $max\{c^T x | x \in P\}$ and let x^* be the optimum solution
$\quad$ **if** there exists a facet defining inequality $a^T x \leq a_0$ such that $a^T x^* > a_0$ **then**
$\quad\quad$ $P := P \cap \{x \in \mathbb{R}^{n(n-1)} | a^T x \leq a_0\}$
$\quad\quad$ $found := true$
$\quad$ **else**
$\quad\quad$ $found := false$
$\quad$ **end if**
end while

if x^* is integral **then**
$\quad$ x^* solves the Linear Ordering Problem
else
$\quad$ start Branch & Bound
end if

---

by applying uniform distribution, the algorithm was not able to solve the problem of size 50, within the time that authors had on their disposal. This result was expected since it reflects a problem instance were almost all possible solutions are close to optimum, which naturally implies a long Branch and Bound search.

3.7 Finding a Minimum Feedback Arc Set in Reducible Flow Graphs [110]—September 1988

The author of the paper presented a polynomial-time algorithm that solves the problem of MFAS for Reducible Flow Graph (RFG). The paper has also established that the problem of FAS, for reducible flow graph with arc weights, is in P. To achieve its goal, and find minimum FAS, the algorithm repeatedly uses the maximum flow algorithm on graphs that are derived from input reducible graph.

Time complexity for the MFAS algorithm in question is $O(mn^2 \log(n^2/m))$, where n represents a number of vertices, and m represents a number of arcs. The author has additionally stated that any algorithm for weighted FAS, or the weighted FVS, on reducible flow graphs has time complexity of at least that of finding a minimum cut in a flow network of the same size—namely $\Theta(mn \log(n^2/m))$.

On the other hand, when we are speaking about unweighted RFGs, the authors algorithm is finding FAS in $O(min(m^2, mn^{5/3}))$ time—in this instance, any algorithm will have a time complexity of at least that of finding a min-cut in a unit-capacity network, namely $\Theta(min(mn^{2/3}, m^{3/2}))$.

The author has also adapted FAS algorithm, and has obtained the algorithm for finding a minimum weight FVS on reducible graph. Time complexity of the weighted problem of FVS is at least that of finding a minimum cut in a flow network. Pseudo-code for the FAS algorithm can be seen in Algorithm 3.7.1—procedure for the actual construction is detailed in the proof of theorem 2.1 of the paper in question.

Algorithm 3.7.1 MFAS for reducible flow graph

Input: A reducible graph $G = (V, A, r)$ with non-negative weights on arcs.
Output: The cost of a minimum FAS for G.

Preprocess G: Label the heads of back arcs in G in postorder. Derive the head dominator tree T_h for G. Introduce a pointer from each vertex i in T_h (except r) to its parent h_i. Let the number of heads be h.
for $i = 1, \ldots, h$ process head i **do**
    find the capacity of minimum cut, c_i, in $G_m(i)$
    if $i \neq h$ then introduce an arc of weight c_i from h_i to i in G {note that G changes during the execution of the algorithm so that $G_m(i)$ is the same as $G_{mm}(i)$ if G were unchanged}
end for
return c_h as cost of minimum FAS for G

3.8   A Contraction Algorithm for Finding Small Cycle Cutsets [94]—December 1988

The paper deals with finding a minimum cut-set in a directed graph. Cut-set is represented by a set of vertices that cut all cycles in $G = (V, E)$. The authors have proposed a new algorithm for finding small cut-sets which works by applying a set of operations, conveniently called by the authors, contraction operations. These operations contract a graph until no contraction is anymore possible, while at the same time the order in which these contractions are done does not affect end result— this property is called finite Church–Rosser.

The algorithm proposed by the authors, namely contraction algorithm, finds small cycle cut-sets by sequentially applying these contractions to directed graph. The algorithm then terminates when an empty contracted graph is reached. Thus, selected vertices form a cut-set of the original graph. Depending on the comparison algorithm in question, a contraction algorithm is either more powerful, or it has broader applicability, or is more efficient, or its improvement over other algorithms is formed by some of the stated improvements—complexity of the contraction algorithm is $O(|E| \log |V|)$. A contraction algorithm skeletal pseudo-code can be seen in Algorithm 3.8.1.

In order to enable algorithm comparison the authors have constructed a hierarchy of classes of graphs. There are two classes of graphs, namely the class of backward contractible graphs and the class of reducible graphs, that are similar and are defined in terms of contraction operations. Authors have found that the class of

---

Algorithm 3.8.1 The contraction algorithm (a skeletal description)

Input: Graph G (containing no parallel edges).
Output: Vertex sets, S_1 and S_2.

(1) If G is the empty graph, stop.
(2) Repeatedly apply the basic contraction operations to G until no longer possible. Collect the vertices removed by the $LOOP$ operation in S_1.
(3) If the graph is empty, stop. Else, select a vertex v, contract the graph by $REMOVE(v)$, add v to S_2, and go to step 1.

{Basic contraction operations can be seen in section 5 of the paper. For a guarantee that the algorithm finds a minimum cut-set of G see propositions 5.1, and 5.3, and corollary 5.4.}

---

quasi-reducible graphs lies in between these two, and that the class of completely reducible graphs, reducible by contraction algorithm, properly contains the class of quasi-reducible graphs, therefore making the contraction algorithm a more powerful option when such matters are of importance. For other related and similar algorithms through which even more detail is revealed you can consult the following literature, [18, 116, 122, 124, 130].

3.9 Approximation Algorithms for the Maximum Acyclic Subgraph Problem [16]—January 1990

In this paper the authors have dealt with MAS, where for a directed graph $G = (V, A)$ the goal is to find a subset A' of the graph arcs such that a subgraph $G' = (V, A')$ is without cycles, and A' is of maximum cardinality.

The authors have presented approximation polynomial time, and RNC (Nicks Class [32] with access to randomization) algorithms which given any graph without two-cycles find a solution of the size at least

$$\left(1/2 + \Omega\left(\frac{1}{\sqrt{\Delta(G)}}\right)\right)|A| \tag{3.2}$$

with $\Delta(G)$ being maximum total degree of V in G—the bound mentioned is tight, namely a class of graphs exists, without two-cycles, for which the largest acyclic subgraph has size at most

$$\left(1/2 + O\left(\frac{1}{\sqrt{\Delta(G)}}\right)\right)|A| \tag{3.3}$$

The authors have presented a linear time $O(|V| + |A|)$ randomized algorithm, $O(|V||A|)$ deterministic algorithm (which was achieved by derandomization of the randomized algorithm, and with the technique of Raghavan and Spencer [109, 125]), and a logarithmic-time, and linear-processor RNC, algorithm of the previous approximation bound. A heuristic for the case of $FAS \leq \frac{1}{2}|A|$, as expounded by [117], can be seen in Algorithm 3.9.1.

Algorithm 3.9.1 Heuristic algorithm for $FAS \leq \frac{1}{2}|A|$

$F := \emptyset$
while $G \neq \emptyset$ **do**
    select a vertex v in G
    if $d^-(v) < d^+(v)$ **then**
        add all arcs incoming to v to F
    else
        add all arcs outgoing from v to F
    end if
    remove v and all arcs incident to it from G
end while
return F

When a graph of a low degree is considered, which the authors state as a common case, algorithms mentioned provided improvement over existing algorithms of the time that were finding a solution of a size $\frac{1}{2}|A|$. The authors state that for graphs without two-cycles, by specific selection of the next vertex to be processed by FAS heuristic, their algorithms find a solution of at least $\frac{2}{3}|A|$, when $\Delta(G) = 2$ or 3; that is $\frac{19}{30}|A|$, when $\Delta(G) = 4$ or 5. The bound is worst case optimal if $\Delta(G) = 2$. And if the algorithms are dealing with 3-regular graphs, then a solution of size at least $\frac{13}{18}|A|$ is achievable, which is slightly better than $\frac{2}{3}|A|$.

3.10 Exact and Heuristic Algorithms for the Weighted Feedback Arc Set Problem: A Special Case of the Skew-Symmetric Quadratic Assignment Problem [54]—January 1990

The author of the paper presents both exact and approximate algorithms for the weighted FAS. In the context in which the author deals with the issue, the problem is that of finding a permutation matrix P where the sum of elements above principal diagonal $P'WP$ is maximized, with W being a skew-symmetric matrix of order n.

To establish computational empiricism the author has researched on a sample of randomly generated problems, including comparisons with three other algorithms for solving a more general quadratic assignment problem. On randomly generated problems computational time that is requited to find all exact solutions for each problem is approx. $T(n) = c2.232^n$, where $c = 9.8764 \times 10^{-6}$; on a Cray X-MP/48 supercomputer $T(20) = 93\ s$. On the other hand, for the authors sample, a computational time that was required to calculate one approximate solution was c. $T(n) = an^{4.1}$, where $a = 3.0361 \times 10^{-8}$; on the Cray $T(250) = 206\ s$.

The algorithm that is working as an exact algorithm is called the screening algorithm. This algorithm is constructing permutations sequentially, and applying screening tests. All solutions are calculated since all possible permutations are considered. The algorithm was tested on randomly generated problems, and calculations were done on an IBM PC with an 8087 coprocessor, an Amdahl 470V/8, and a Cray X-MP/48. For the use on an IBM PC the algorithm was coded once in IBM APL, and once in MICROSOFT Fortran 4.0—Fortran code was used under MTS on the Amdahl.

To take advantage of vectorization, and other features of Cray supercomputer, Fortran code was customized for the Cray. The screening algorithm was compared to the algorithm of Edwards [48], and the results showed that the screening algorithm performed significantly better than the Branch and Bound algorithm of Edwards—for special cases of the Koopmans-Beckmann problem. All calculations were performed on an IBM PC computer, for more details see table 1. of the paper in question. A somewhat simpler screening algorithm was compared to Burkard-Derigs Fortran program [24]. Even in this simpler version of the algorithm problems of order greater than 20 are difficult to solve. But compared with Burkard-Derigs program the simpler screening algorithm performs well.

Screening algorithm was for higher order instances of the problem too computationally intensive, therefore an algorithm that yields good approximate solutions but is more efficient was clearly a good alternative. Approximation algorithm that was developed by the author yields in most instances exact solutions (dependent on number of repetitions), while at the same time this algorithm needs much less computational effort than the screening algorithm needed. Computational results for the approximation algorithm are limited, and illustrative—results were promising, and can be seen in table 4. and 5. of the paper in question. Largest instances analyzed were of order 300, and required c. 8 min on a Cray X-MP/48 supercomputer. When compared to the algorithm by Kaku et al. [79], solution wise the algorithm of Flood fares well with computational time being much less.

3.11 Leighton-Rao Might Be Practical: Faster Approximation Algorithms for Concurrent Flow with Uniform Capacities [84]—April 1990

In this paper authors have presented new algorithms, that are approximate in nature, for the problem known as concurrent multi-commodity flow with uniform capacities. The problem of multi-commodity flow deals with shipping different commodities, each from its source, to their destination in a single network. Total amount of flow is going through edges that are limited by their capacity. The authors are in their paper considering the concurrent flow problem with uniform capacities.

To be exact, they are trying to find a multi-commodity flow that minimizes congestion, that is a total flow on any edge.

There is another version of the problem, which the authors have also tackled, where the amount of shipped flow is same for each commodity—this problem is a special case, and is known under a name unit-demand and unit-capacity concurrent flow problem.

Leighton and Rao [92] used concurrent flow to find approximately "sparsest cut" in a graph. In this way they have in an approximate way solved a number of graph problems, including the one that is important to us, namely MFAS. Using linear programming techniques they devised polylog-time-optimal approximation algorithms for a number of NP-Complete graph problems, with computational bottleneck being solving a unit-capacity and unit-demand concurrent flow with $O(n)$ commodities. By using the method of Leighton and Rao the authors of the paper in question have showed that method of Leighton and Rao might have practical value, and have given an $O(m^2 \log m)$ expected-time randomized algorithm—this algorithm solves concurrent flow problem on an m-edge graph. Algorithms devised by the authors find a solution, for any positive e, whose congestion is $\leq (1+e)$ times the maximum congestion.

Procedure called by the authors REDUCE takes as input a multi-commodity flow f, a target value τ, an error parameter e and a flow quantum σ. After this procedure finishes execution the output will be an improved multi-commodity flow f, where either $|f|$ is less than the target value τ, or f is e-optimal. Pseudo-code for the procedure can be seen in Algorithm 3.11.1.

Algorithm 3.11.1 Reduce(f, τ, e, σ)

$\alpha \leftarrow 10\tau^{-1}e^{-1}\log(9me^{-1})$
while $|f| \geq \tau$ and f and l are not e-optimal, for each edge vw, $l(vw) \leftarrow e^{\alpha f(vw)}$ **do**
    $FindPath(f, l, e)$ {so as to find an e-bad path P, and a short path Q with the same endpoints as P}
    reroute σ units of flow from P to Q
end while
return f

As stated in lemma 4.5 of the paper in question, with description of the iteration in prior text, total time for REDUCE when FindPath is implemented in a deterministic manner is $O(min\{n, k\}(m + n\log n) + m\log n)$. When e is not very small a randomized implementation can be achieved that is much faster. This Monte-Carlo implementation can be achieved in an expected time of $(e^{-1}(m + n\log n))$. To make the algorithm Las Vegas, a deterministic check to determine whether f and l are e-optimal can be introduced, and in turn the total expected contribution to the running time of REDUCE is $O(k(m + n\log n))$. The authors have also showed that if one would be satisfied with approximate shortest path granularity (small enough flow quantum to guarantee that approximate optimality is achievable) then REDUCE takes expected time of $O(\frac{D}{\sigma}m)$ for a randomized version of FindPath, and can be implemented in $O(\frac{D}{\sigma}m(\log n + min\{n, k\}))$ time for a deterministic version of FindPath.

For a special case of concurrent flow problem authors have devised two approximation algorithms. This special case is called unit-capacity and unit-demand concurrent flow problem, where capacities are uniform and the demand $d(i)$ is the same for each of the k commodities. First of these two algorithms is called Uniform, which is simpler and works best if e is a constant value. The algorithm is finding an e-optimal multi-commodity flow in expected time of $O((ke^{-2}\log k + me^{-4}\log m)(m + n\log n))$, and if $0 < e \leq 1$ an e-optimal solution can be found in expected time of $O(m(k\log k + m\log m))$. For pseudo-code of Uniform consult Algorithm 3.11.2.

The other algorithm is called ScalingUniform, which starts with a large e and then gradually scales e down—this algorithm is faster for small values of e. The scaling is achieved by calling Uniform with $e = 1$, and then repeatedly dividing e by 2 and calling REDUCE. This algorithm is finding an e-optimal solution in expected time of $O((ke^{-1}+k\log k+me^{-3}\log m)(m+n\log n))$, and if REDUCE is deterministic then a solution for unit-capacity and unit-demand concurrent flow can be obtained in time $O((k\log k + me^{-2}\log m)(m\log n + min\{n, k\}(m+n\log n)))$. For pseudo-code of ScalingUniform you can consult Algorithm 3.11.3.

Algorithm 3.11.2 Uniform($G,\ e,\ d,\ \{(s_i,\ t_i) : 1 \leq i \leq k\}$)

---

for each commodity i, create a simple path from s_i to t_i and route d flow on it
$\sigma \leftarrow d$ and $\tau \leftarrow \frac{|f|}{2}$

Phase U1
while σ and τ satisfy the granularity condition and f is not e-optimal **do**
    $Reduce(f,\ \tau,\ e,\ d)$
    $\tau \leftarrow \frac{\tau}{2}$
end while

Phase U2
until σ and τ satisfy the granularity condition, $\sigma \leftarrow \frac{\sigma}{2}$
while f is not e-optimal **do**
    $Reduce(f,\ \tau,\ e,\ \sigma)$
    $\sigma \leftarrow \frac{\sigma}{2}$ and $\tau \leftarrow \frac{\tau}{2}$
end while

return f

At the end of algorithmic part of the paper the authors have presented a randomized procedure GENERAL which is finding approximate optimal solutions for the unit-capacity concurrent flow problem where demand for every commodity is being rounded to an integer multiple. Pseudo-code of this procedure can be seen in Algorithm 3.11.4. This algorithm finds $3e$-optimal flow solution with an expected time of $O((mke^{-3} + me^{-4}\log m)(m + n\log n))$. For possible alternate algorithm

---

Algorithm 3.11.3 ScalingUniform(G, e', d, $\{(s_i,\ t_i) : 1 \leq i \leq k\}$)

$e \leftarrow 1$
$Uniform(G,\ e,\ d,\ \{(s_i,\ t_i) : 1 \leq i \leq k\})$, and let f be the resulting flow
$\tau \leftarrow \frac{\tau}{2}$
if $\sigma < d$ **then** go to phase S2

Phase S1
while $e > e'$, and σ and τ satisfy the granularity condition, $e \leftarrow \frac{e}{2}$ **do**
  $Reduce(f,\ \tau,\ e,\ d)$
end while

Phase S2
while $e > e'$, $e \leftarrow \frac{e}{2}$, $\sigma \leftarrow \frac{\sigma}{4}$ **do**
  $Reduce(f,\ \tau,\ e,\ \sigma)$
end while

return f

---

versions, and subsequent theorems please look at theorems 6.3–6.5 of the paper in question.

For applications that authors had in mind you can look at section 7 of the paper in question, Two Applications, where the authors go into more detail on Leighton and Rao approximation algorithm implementation, and approximately minimizing width of the channel in VLSI so as to minimize the total circuit area.

Algorithm 3.11.4 General(G, e, $d(i)$, $\{(s_i,\ t_i) : 1 \leq i \leq k\}$)

$\tau \leftarrow D$, and $\sigma \leftarrow min \left\{ \frac{e\tau}{k},\ \frac{e^2\tau}{200\log(9me^{-1})} \right\}$

for each commodity i, find a simple path from s_i to t_i and route $d^\sigma(i)$ flow on it

while f is not e-optimal **do**
  $Reduce(f,\ \tau,\ e,\ \sigma)$
  $\sigma \leftarrow \frac{\sigma}{2}$ and $\tau \leftarrow \frac{\tau}{2}$
  for each commodity i, find a simple path from s_i to t_i and route $d^\sigma(i) - d^{2\sigma}(i)$ flow on it
end while

for each commodity i, find a simple path from s_i to t_i and route $d(i) - d^\sigma(i)$ flow on it
return f

3.12   An Efficient Method for Finding a Minimal Feedback Arc Set in Directed Graphs [105]—May 1992

To tackle the problem of MFAS the authors have firstly partitioned a graph into strongly connected and bi-connected components. Then, using a well known Depth-First Search, each component is traversed. The ordering is done in such a way where only backward arcs are negative arcs. When this state is achieved, an additional efficient cutting technique is employed so as to reduce the negative arcs further. When compared to existing random sequential ordering method [133] the proposed algorithm shows substantial improvement.

To shorten the search for MFAS of a graph G various widely used simplifications can be applied: self-loop removal (single vertex and a self-arc), source/sink removal (no incoming/outgoing arc exists for a vertex), strongly connected component partitioning (for every vertex u and v directed path from u to v exists), and bi-connected component partitioning (containing articulation point causing graph disconnection).

The entire algorithm that the authors have devised have been implemented in C language, and was run on the VAX environment. Directed graphs that were used in a benchmark were obtained by sequential circuits [22], where each gate was changed into a vertex and each lead into an arc.

Generally speaking, by comparing authors DFS to random sequential ordering, DFS makes a reduction that is at least 50% in number of negative arcs—this reduction can stretch all the way up to 90%. After that, by applying a cutting technique, further large reductions can be made—this technique works well for large strongly connected components.

3.13   A Fast and Effective Heuristic for the Feedback Arc Set Problem [47]—October 1993

In this work a greedy heuristic for FAS with a good non-optimal performance bound is presented, the algorithm executes in linear $O(m)$ time, and also $O(m)$ space. When sparse graphs are considered, which arise, for example, in graph drawing, the algorithm is achieving the same asymptotic performance bound as Berger and Shor [16] algorithm does which has a running time of $O(mn)$.

If one would need an algorithm for an online processing, then a faster algorithm would be preferred, and therefore the authors have devised such an algorithm with a suboptimal but efficient performance. The algorithm greedily removes vertices, and incident arcs, from G which are a sink or a source. Algorithm pseudo-code can be seen in Algorithm 3.13.1—for a more elegant version one can consult [117].

Algorithm 3.13.1 outputs either an empty sequence of vertices, or a sequence s for which $|R(s)| \leq m/2 - n/6$. If one deals with tournaments then a sharper bound

---

Algorithm 3.13.1 GR($G : DiGraph$; var $s : VertexSequence$)

$s_1 \leftarrow 0; s_2 \leftarrow 0$
while $G \neq 0$ **do**
    while G contains a sink **do**
        choose a sink u; $s_2 \leftarrow us_2$; $G \leftarrow G - u$
    end while
    while G contains a source **do**
        choose a source u; $s_1 \leftarrow s_1u$; $G \leftarrow G - u$
    end while
    choose a vertex u for which $\delta(u)$ is a maximum
    $s_1 \leftarrow s_1u$; $G \leftarrow G - u$
end while
return $s \leftarrow s_1s_2$

---

can be formulated, that is, algorithm GR computes a sequence of vertices s where $|R(S)| \leq m/2 - \lfloor n/2 \rfloor /2$.

For the algorithm implementation it is convenient to partition vertices into: sources, sinks, and δ-classes (difference between out-degree and in-degree); where every vertex $u \in V$ belongs only to one of $2n - 3$ classes. Therefore to compute these classes a bin sort can be executed, as initialization. In this way each vertex class is represented as a bin with vertices in classes linked by a doubly linked list.

3.14 Approximations for the Maximum Acyclic Subgraph Problem [69]—August 1994

This paper deals with a well known problem of Maximum Acyclic Subgraph, a dual of Minimum Feedback Arc Set. The problem belongs to an edge deletion set of problems. The main contribution that the authors have given is an algorithm for the unweighted version guaranteeing a bound similar to that of Berger et al. [16] with time complexity of $O(|A| + d_{max}^3)$—which in some cases represents a better bound than $O(|A| \cdot |V|)$. The algorithm can be seen in Algorithms 3.14.1 and 3.14.2. Procedure 3.14.2 has an expected number of arcs in a solution equal to

$$\left(1/2 + \Omega\left(\frac{1}{\sqrt{d_{max}}}\right)\right)|A| \tag{3.4}$$

This procedure can be derandomized, turned into a deterministic one, by method of conditional probabilities [5] while at the same time retaining performance guarantee.

As for the procedure in Algorithm 3.14.1 there are a number of possible selection rules that can be applied to examination of vertices, where one of the possibilities would be selection as per Algorithm 3.14.3. It is naturally also possible, as have the authors also stated, to search by making a change only on one vertex, and

---

Algorithm 3.14.1 Permutation generating procedure

Step 1
$S = V, l = 1, u = n$

Step 2
choose $i \in S$
$S \leftarrow S \setminus \{i\}$
if $w_i^{in}(S) \leq w_i^{out}(S)$ then
   $\pi(i) = l$
   $l \leftarrow l + 1$
end if
if $w_i^{in}(S) > w_i^{out}(S)$ then
   $\pi(i) = u$
   $u \leftarrow u - 1$
end if

Step 3
if $u \geq l$ then
   go to step 2
else
   return π
end if

Algorithm 3.14.2 Randomized approximation procedure for MAS

Step 1
partition V into two subsets V_1, V_2 by assigning each vertex to each subset with probability $1/2$

let $A_r = \{(i, j) \mid i, j \in V_r\}$
execute step 2 for $r = 1, 2$

Step 2
form a permutation π, of the vertices in V_r by applying Algorithm 3.14.1 to (V_r, A_r)
vertices are selected in increasing order of their indices

Step 3
define the final permutation by choosing between (π_1, π_2) and (π_2, π_1) the permutation inducing a subgraph with the larger number of arcs

---

thus searching locally by examining whether a better permutation exists in a neighborhood.

3.15   A Heuristic for the Feedback Arc Set Problem [46]—September 1995

The authors of the paper have presented a heuristic algorithm for FAS which they named FASH. Their algorithm has refined results from [47] so as to reduce the size

---

Algorithm 3.14.3 An alternate approach to permutation construction

---

Step 1

choose $(i, j) \in A$

set $Q = \{i, j\}, i \prec j, S = V \setminus Q$

Step 2

suppose that this step is reached with an order $\prec$ of Q, choose $i \in S$ and set $S \leftarrow S \setminus i$

compute for each $j \in Q$, $D_j = \sum_{k|k \leq j} w_{ki} + \sum_{k|j \prec k} w_{ik}$

set $D_0 = \sum_{j \in Q} w_{ij}$

let $D_l = max\{D_j \mid j \in Q\}$

if $D_l \geq D_0$ **then**

    extend $\prec$ by adding i so that it is the immediate successor of l

end if

if $D_l < D_0$ **then**

    extend $\prec$ by adding i as the first element

end if

Step 3

if $S = \emptyset$ **then**

    stop

else

    set $Q \leftarrow Q \cup \{i\}$

    go to step 2

end if

---

of resultant feedback arc set. The algorithm is computing a vertex ordering for a digraph, and is then outputting a set of leftward arcs as a feedback arc set. The algorithm consists of the following steps:

1. **Iteratively remove sinks** (if any) to prepend to a vertex sequence s_2; and if the remaining graph is empty, then go to step 4, else go to step 2.
2. **Iteratively remove sources** (if any) to append to a vertex sequence s_1; and if the remaining graph is empty, then go to step 4, else go to step 3.
3. **Choose a vertex** u, such that the difference between the number of rightward arcs and the number of leftward arcs is the largest, and remove u to append to s_1; if the remaining graph is empty, then go to step 4, else go to step 1.
4. **Vertex sequence** s **is formed** by concatenating s_1 with s_2; and the leftward arc set for the vertex sequence s is reported as a feedback arc set.

Steps 1 and 2 do not produce any feedback arcs, and if there exists more than one candidate in step 3, then the algorithm in [47] chooses one of the options in a nondeterministic manner. By step 3 is therefore execution of the algorithm potentially sped up, but in this way it can also happen that performance is diminished. The authors of the new heuristic algorithm have thus focused their effort on this step, so as to improve the situation—they have added additional greedy criteria for choosing a vertex, as well as done manipulations for a digraph to allow for this step to be more effective.

When considering structure of a digraph, graph can be decomposed into strongly connected components—this can be done in linear time. In this way one can obtain components of a digraph G with a property where there are no leftward arcs between components. Therefore one only needs to find FAS for individual components.

Algorithm 3.15.1 FASH(G : *directed graph*) : *vertex sequence*

$s \leftarrow \emptyset$
$(G_1, G_2, \ldots, G_k) \leftarrow DSC(G)$
return $SCFASH(G_1) \cup SCFASH(G_2) \cdots \cup SCFASH(G_k)$

Such digraph components might be condensible. That is, a directed path $(u_1, u_2, \ldots, u_k)$ is condensible if $k \geq 3$, $d_G(u_1) \geq 3$, $d_G(u_k) \geq 3$, and for $2 \leq i \leq k - 1$, $d_G^+(u_i) = d_G^-(u_i) = 1$; u_1 and u_k are, respectively, called start and end vertex of a digraph, while u_i represents middle vertex of a digraph. A digraph G is completely condensed if it is strongly connected, and there are no directed paths that can be condensed—condensation is performed by a path collapsing to a single arc.

When a digraph is fully condensed it would be opportune to choose a vertex v where the remaining graph after deletion of v contains a vertex u which is "unbalanced," that is there is a high difference between in and out-degree of u. In this way the algorithm will potentially in its next iteration produce a small number of feedback arcs. Algorithm of the authors can be seen in Algorithms 3.15.1, 3.15.2 and 3.15.3.

Algorithm 3.15.2 SCFASH(G : *strongly connected graph*) : *vertex sequence*

$s \leftarrow \emptyset$
$G_c \leftarrow CON(G)$
$v \leftarrow OBTAIN(G)$
return sequence formed by prepending v to $FASH(G - v)$

Procedure DSC is returning strongly connected components of a digraph G where there are no leftward arcs between components. Procedure CON computes condensation G_c of a strongly connected digraph G. Procedure TAKEMAX outputs a set of vertices U for a digraph G with maximum value of difference between in and out-degree. And lastly, procedure CHS produces extremely "unbalanced" vertices, and returns a vertex v in U for a graph G.

All procedures can be implemented in linear time, algorithm FASH executes in $O(mn)$ time, where n represents a number of vertices, and m represents a number of arcs. If the algorithm is restricted to cubic digraphs, with no two-cycles and no loops, it produces FAS of cardinality $\leq \frac{m}{4}$, where m is a number of arcs in G. For a general simple digraph the reported performance bound was $\frac{m}{2} - \frac{n}{6}$.

Algorithm 3.15.3 OBTAIN(G : *graphs*) : *vertex*

---

if G has only one vertex v **then**
   return v
else
   $TAKEMAX(G, U)$
   return $CHS(G, U)$
end if

3.16 New Results on the Computation of Median Orders [26]—March 1997

Here the issue is, given a weighted tournament T, finding a set of arcs in T, of minimum weight, such that when one reverses these arcs it makes T transitive. This problem is quite closely connected to the problem of FAS, and is in fact a generalization of the same problem. To tackle the aforementioned issue the authors have made improvements to the Branch and Bound type of algorithm.

A median order, median linear order, or minimum reversing set, is a linear order minimizing total weight of the arcs which do not have the same orientation as in T—where $T = (X, U)$ is a tournament weighted by w.

The evaluation function of the initial Branch and Bound algorithm [14] has been improved by anticipating the remoteness of a tournament, where remoteness r represents a measure over the set of linear orders that is to be minimized. This evaluation function E has been improved by the addition of three parameters. The first of these is $\chi(T)$, where $C_3(T)$ is the number of 3-circuits of T and $(\delta(u_j))_{1 \leq j \leq n_H}$ denotes decreasing degrees of the hypergraph H. The parameter $\chi(T)$ is the least integer χ where $\sum_{j=1}^{\chi} \delta(u_j) \geq C_3(T)$. Therefore instead of the initial Branch and Bound function E, a new one may be considered. Let $T_{\bar{N}}$ be sub-tournament induced by vertices which do not belong to N, then $E(N) + W_\chi(T_{\bar{N}})$ is a lower bound of the remoteness between T and any linear order beginning by N. $E(N)$ is a measure of the weights of the arcs that must be reversed to get N, and $W_\chi(T_{\bar{N}})$ is a measure of anticipation for what will be necessary to reverse in $T_{\bar{N}}$ so as to complete N into a linear order defined on the whole set X (of vertices of T).

The second of these parameters is connected to the maximum number $\zeta(T)$ of arc-disjoint 3-circuits of T. In order to avoid looking for a maximum-weighted set of arc-disjoint circuits of T one can restrict himself to 3-circuits of T and try to find only a lower bound of the maximum weight of a family of arc-disjoint 3-circuits. Therefore a computation can be made for all 3-circuits C of T and their values $\mu(C)$ (minimum of weights of arcs belonging to C) before starting Branch and Bound algorithm. Consequently, for each node N of the Branch and Bound tree, one can greedily compute (as heavy as possible) family of arc-disjoint 3-circuits of the sub-tournament $T_{\bar{N}}$. Thus by setting $W_\mu(T_{\bar{N}}) = \mu(F(T_{\bar{N}}))$ a lower bound of $r(T_{\bar{N}})$ is obtained—where weight $\mu(F) = \sum_{j=1}^{k} \mu(C_j)$.

The third parameter deals with the scores of vertices of T, where $s_1 \leq s_2 \leq \cdots \leq s_n$ denote (increasing) scores of vertices of T, and the parameter is then defined by

$\sigma(T) = \frac{1}{2} \sum_{j=1}^{n} |s_j - j + 1|$. Parameter $\sigma(T)$ can be used to improve evaluation of a node N of the search tree by adding the weights of $\sigma(T)$ least weighted arcs of T: $W_\sigma(T) = \sum_{j=1}^{\sigma(T)} w(u_j)$, the arcs of T being ranked by increasing weights, thus $W_\sigma(T) \leq r(T)$.

Therefore an evaluation function E can be defined. For any node N of the branch and bound tree, $E'(N) = E(N) + Max\{W_\chi(T_{\bar{N}}), W_\mu(T_{\bar{N}}), W_\sigma(T_{\bar{N}})\}$ is a lower bound of the remoteness between T and any linear order with N as a beginning section. It is also possible to eliminate certain beginning sections if another beginning section is known and is defined on the same set of vertices, but in some other order and with strictly lower evaluation E.

Before the paper is concluded, three necessary conditions for a vertex to be a winner are given (x is said to be a winner of T if there exists a median order whose first vertex is x), to these conditions an additional condition needs to be added where it is imposed for a winner to be the first vertex of a Hamiltonian path of T. For a detailed look one can consult propositions 4.2–4.4 of the paper in question.

Unlike in the original Branch and Bound algorithm where a best-first expanding strategy was applied, here the authors have opted for a Depth-First strategy. The former strategy usually has less nodes, and the first linear order computed on X is median. The downside is that memory consumption is much more pronounced than in the Depth-First strategy. Best-first strategy requires keeping of all the nodes of the best-first tree, while Depth-First requires to keep only the current branch that is being expanded in the tree. In a best-first situation one also needs to find the leaf with the lowest evaluation, which may be very time consuming, even with an appropriate data structure such as the heap, since Branch and Bound tree may increase exponentially with n. Lastly, with a Depth-First strategy it was easier to update information coming from one level of the tree to the next one. This can be done in $O(1)$ time for the Depth-First strategy and the improved algorithm, while it would take an $O(n)$ time for the initial algorithm to compute from nothing.

Initial algorithm could be adapted to store more information in the nodes of the tree, and in this way reduce computation, but that would expand even further the problem of memory consumption. There are six properties through which the authors have tried to reduce the computing time, and size of the search tree. They are as follows (references in brackets are for readers convenience and are referring to the paper in question):

1. Hamiltonian path (Corollary 2.2)
2. Improvements of the evaluation function, **Evaluation Function** (Proposition 3.4)
3. Elimination of some beginning sections defined on the same set of vertices, **Beginning Sections**
4. Source, **Source Property** (Propositions 4.1 and 4.2)
5. Uncovered set, **UnCovered** (Proposition 4.3)
6. Weight, **Weight Property** (Proposition 4.4).

During experimentation the authors have constantly used Hamiltonian path property, this was always used in the initial Branch and Bound algorithm as well, but other properties were optional and the authors were examining different combinations of these so as to ascertain which one is the most interesting. Since complexity of different properties is not the same it was not possible to ascertain them a priori. Experiment was composed of 25 random tournaments were arcs (i, j) and (j, i) were being chosen uniformly. For weighted tournaments weights of the arcs were chosen randomly and uniformly over the range of integers [0, 9]. While in the unweighted tournaments, all weights are equal to 1.

Conclusions from many tests with different values of n are therefore as follows. For weighted tournaments the most powerful properties are WP and BS. These two ways of cutting do not affect the same nodes of the Branch and Bound tree, and combining them gives a result similar to what is achieved when all the properties are active, although in a much shorter time. Property EF is for weighted tournaments not very efficient due to necessary computing time in order to find lowest-weighted arcs and to compute the sum of their weights. Properties UC and SP usually do not bring outstanding savings for the size of the tree, while on the other hand there is the time required to compute them.

Unweighted tournaments bring slightly different conclusions. Here the most useful property is EF, and best combination of properties is represented by combining EF and WP. They are considerably cutting the search tree, and consequently bringing substantial savings in terms of CPU time. These two alone are producing almost the same results as when all the properties are used. When WP is used standalone then the results are not that impressive, since random generation produces tournaments with almost equal scores, and in such a situation WP cannot cut many nodes, although it is quick to compute. Property BS is for unweighted tournament not of much interest. UC and SP are here as well not efficient, the culprit is random generation, as is with weighted tournaments. For tournaments close to the ones that are transitive, SP may sometimes have its use, since it does not consume much time to be applied.

3.17 Approximating Minimum Feedback Sets and Multi-cuts in Directed Graphs [50]—February 1998

Authors of this paper have worked both on weighted feedback vertex set (FVS) problem, and weighted feedback edge set (FES) problem. Generalizations of these two problems have also been considered, namely Subset-FVS and Subset-FES—in which the feedback set has to intersect only a subset of the directed cycles in the graph. Such a subset consists of all cycles which go through a distinguished input subset of vertices and edges, and even if number of vertices equals 2 this generalized version is still NP-Hard. If a number of vertices is 1, then the problems can be solved efficiently, that is in polynomial time by computing a min-cut.

In a separate section of the paper the authors have presented different versions of problems tackled, and reductions among these problems. All reductions map feasible solutions, and the cost is preserved, therefore an approximate solution to one problem can be translated to approximate solution to other problem. Of these reductions, most can be performed in linear time. Reductions performed are as follows: FES $\preceq$ FVS, FVS $\preceq$ FES, Blackout-FVS $\preceq$ FVS, Subset-FES $\preceq$ directed multi-cuts. Blackout-FVS is an extension of FVS where an additional subset of "blackout" vertices is given. Thus only feedback vertex sets that do not contain any "blackout" vertices are allowed.

For these generalized versions of the problems, Subset-FVS and Subset-FES, authors have devised polynomial-time approximation algorithms. As a prestep, a polynomial-time approximation algorithm for the multi-cut problem in circular networks has been presented. This algorithm has an approximation factor of $O(log^2 k)$, where k is the number of source-sink pairs. The algorithm is based upon a decomposition of the graph, and an adaptation of the undirected "sphere growing" technique to directed circular networks, and is applied to weighted Subset-FVS and weighted Subset-FES so as to obtain aforementioned approximation. Next, a polynomial-time approximation algorithm for the weighted generalized feedback problems is presented. This algorithm is finding a feedback set with weight $O(min\{\tau^* \log \tau^* \log \log \tau^*, \tau^* \log n \log \log n\})$, where τ^* represents a cost of an optimal fractional feedback set, and n is the number of vertices.

The paper also presents a combinatorial algorithm for the fractional FVS and Subset-FVS problems that is finding a $(1 + \epsilon)$ approximation. The complexity of this particular algorithm is $O(n^2 M(n) \log^2 n)$ for any fixed ϵ, where $M(n)$ denotes complexity of matrix multiplication. A greedy approximation algorithm for the Set Cover problem, derived from parallel algorithm for approximating positive linear programming [96], has been successfully adapted to FVS and Subset-FVS, in spite of exponential number of constraints in the corresponding Set Cover problem. A pseudo-code for this integral multi-cover approximation algorithm can been seen in Algorithm 3.17.1, where V denotes a universal set, F represents a collection of subsets of elements from V, and $c(\cdot)$ is a non-negative cost function associated with each element of V. Detailed description of how to implement this procedure in polynomial time can be seen in section 5.2 of the paper in question. For a more efficient, $O(m \cdot n)$, but a more complex algorithm one can look at [49].

3.18 A Fast and Effective Algorithm for the Feedback Arc Set Problem [117]—May 2001

The author has presented a divide-and-conquer approach as a means of solving FAS. For a step that is dividing the problem a Minimum Bisection (MB) problem is being solved. In order to solve MB, two strategies have been employed: a heuristic based on Stochastic Evolution (SE), and a heuristic based on Dynamic Clustering (DC).

In order to develop a better algorithm, and more precisely, a better division, motivated by the work of Eades, Smyth, and Lin, the author has asked: Which arcs

---

Algorithm 3.17.1 Approximate_SC($V, F, c(\cdot), \epsilon$)

---
if $\epsilon > 1$ **then**
    $\epsilon = 1$
end if
$\lambda_1 \overset{\triangle}{=} 1 + \epsilon/4, \lambda_2 = 4/\epsilon$
$\forall v \in V : l(v) = 0$
$\forall S \in F : p(S) = 1$
repeat
    choose an element in $\{v \in V : p(v) > m^{-\lambda_2}\}$ that minimizes the ratio $c(v)/p(v)$
    $l(v) \leftarrow l(v) + 1$
    $\forall S$ such that $v \in S : p(S) \leftarrow p(s)/\lambda_1$
until $p(v) \leq m^{-\lambda_2}$ for every $v \in V$
return $(\{l(v) : v \in V\})$

---

are forced into the feedback arc set by the divide step? In a divide-and-conquer heuristic by Eades et al. graph G is divided into two sub-graphs G_1 and G_2, where in the final ordering vertices of G_1 precede those that are in G_2. Feedback arc set clearly consists of all leftward arcs with respect to this ordering—arcs forced into the feedback arc set are those originating in G_2 and terminating in G_1. Therefore in order to deliver small feedback arc sets, divide step needs to minimize number of arcs that become leftward, after each step.

In order to achieve efficiency one must find a balanced partition of a graph. For a graph $G(V, E)$ one could define bisection through partition (V_1, V_2) of V where cardinalities of the two subsets (V_1 and V_2) are as close as possible. This can be achieved by requiring $|V_1| \leq \alpha|V|$ and $|V_2| \leq \alpha|V|$, where $1/2 \leq \alpha < 1$ is a constant. Graph Bisection (GB) is a generalization of the Undirected Graph Bisection (UGB) which is known to be NP-Hard [59]. This problem of an unlikely efficient polynomial-time algorithm for GB can be accommodated by fast and effective heuristic. If one would have a function that takes as input $G(V, E)$ and returns a bisection (V_1, V_2) of G of a small cost, an approach for FAS could be defined as is presented in Algorithm 3.18.1. Such a function is used in a strategy for a divide-and-conquer, and is aiming to produce solution of a large graph by the solution of smaller graphs (an arc can be part of a cycle only if its origin and terminus are in the same strongly connected component of the input graph). In order to decompose the graph into smaller pieces, an algorithm that partitions vertex set into several subsets, each of which induces one strongly connected component, was used [35].

In the aforementioned algorithm, scc(G) is a function that takes as input a digraph $G = (V, E)$, and returns a partition $P = \{S : S \subseteq V, S$ induces strongly connected component of $G\}$. Function FAS returns a FAS F of its input graph. If G is not strongly connected, then F is computed as the union of sets where each is a FAS in one strongly connected component of G. If that is not the case, if G is strongly connected, then a function bisect is used to decompose the vertex set of G, into subsets V_1 and V_2 that are about equal in size, after which set F is computed.

Algorithm 3.18.1 FAS(G)

$P := scc(G)$
if P has only one element $\{G$ is strongly connected$\}$ **then**
    $(V_1, V_2) := bisect(G)$
    $F := FAS(G[V_1]) \cup FAS(G[V_2]) \cup \{i \rightarrow j : i \in V_2, j \in V_1\}$
else
    $F := \emptyset$
    for all $S \in P$ **do**
        $F := F \cup FAS(G[S])$
    end for
end if
return F

Stochastic evolution refers to a process where an initial solution is iteratively improved through a sequence of local perturbations. In this way partitions V_1 and V_2 can be changed by moving vertices among them, and thus one can incrementally acquire a more suitable solution. Algorithm 3.18.2 describes perturbations, moves, and gains. The function additionally uses parameter $p \leq 0$ and two stacks S_1 and S_2 to store, in last-first order, vertices moved from V_2 to V_1 and from V_1 to V_2, respectively.

Algorithm 3.18.2 perturb(V, V_1, V_2, p)

$S_1 := \emptyset$
$S_2 := \emptyset$
for all $i \in V$ **do**
    if $gain(i) > randint(p, 0)$ **then**
        $move(i)$
        if $i \in V_1$ **then**
            push i onto stack S_1
        else
            push i onto stack S_2
        end if
    end if
end for
if $|V_1| > |V_2|$ **then**
    $j = 1$
else
    $j = 2$
end if
while $|V_j| > \alpha|V|$ **do**
    pop a vertex i from S_j
    $move(i)$ $\{$reverses previous $move(i)\}$
end while

Function $move(i)$ transfers vertex i from its current subset to the complementary subset, in the partition (V_1, V_2). And function $gain(i)$ represents a reduction in cost, $cost(V_1, V_2)$, after the execution of $move(i)$. If $i \in V_1 : gain(i) = cost(V_1, V_2) -$

$cost(V_1-\{i\}, V_2\cup\{i\})$. If $i \in V_2 : gain(i) = cost(V_1, V_2)-cost(V_1\cup\{i\}, V_2-\{i\})$. Lastly, function $randint(l, h)$ is returning a random number from the same interval. With dependencies defined, the overall Stochastic Evolution algorithm for GB can now be completed, and is given in Algorithm 3.18.3.

Algorithm 3.18.3 Overall SE for GB

$(V_1, V_2) :=$ a random initial bisection of $G(V, E)$
$(B_1, B_2) := (V_1, V_2)$ {save best bisection}
$p := p_0$ {initial value for parameter p}
set value for iteration control parameter R
$counter := 0$

repeat
$\quad C_{pre} := cost(V_1, V_2)$
$\quad perturb(V, V_1, V_2, p)$
$\quad C_{post} := cost(V_1, V_2)$
$\quad$ **if** $C_{post} < C_{pre}$ **then**
$\quad\quad (B_1, B_2) := (V_1, V_2)$ {save best bisection}
$\quad\quad counter := counter - R$ {allow for more iterations}
$\quad$ **else**
$\quad\quad counter := counter + 1$
$\quad$ **end if**
$\quad$ **if** $C_{post} = C_{pre}$ **then**
$\quad\quad p := p - \delta$ {decrease p to allow for more movements of vertices in perturb}
$\quad$ **else**
$\quad\quad p := p_0$ {restore original value of p}
$\quad$ **end if**
until $counter > R$
return (B_1, B_2)

Algorithm 3.18.3 needs three parameters (R, p_0, δ) where R controls the number of iterations, p_0 initializes the variable p which in turn controls the rate of vertex movements in perturb, and δ decreases parameter p so as to signify that little or no change in the current bisection has occurred. If the cost of the current bisection has changed, then p is reset to its initial value p_0. In the authors implementation of SE the following values of the parameters where used: $R = 10$, $p_= -1$, and $\delta = 2$. Out of all iterations, the best bisection that was found is in the end returned.

As per authors definition, clustering, contraction, or compaction describe an operation in which a subset X of vertices of a graph $G(V, E)$ is coalesced and forms a single new vertex x. And therefore arcs entering/leaving a vertex which was originally in X, enter/leave the new vertex x in the modified graph. If clustering is combined with iterative improvement, as was done in [37, 118], then such a strategy can be called Dynamic Clustering. A pseudo-code for the Dynamic Clustering algorithm for GB is described in Algorithm 3.18.4.

Function improve_and_cluster($H, P_1, P_2, H', P_1', P_2'$), of an Algorithm 3.18.4, takes as its input a graph $G(V, E)$, it generates an initial bisection (V_1, V_2) of G, and outputs a compacted graph $G'(V', E')$ along with a bisection (V_1', V_2') of G'.

Algorithm 3.18.4 Dynamic clustering for GB

---
generate a random initial bisection (V_1, V_2) of G
save (V_1, V_2) as the best current bisection

while improvements are made **do**
   let H be a copy of graph G
   let (P_1, P_2) be the best current bisection
   while improvement or compactions are made **do**
      $improve\_and\_cluster(H, P_1, P_2, H', P_1', P_2')$
      $H := H'$
      $(P_1, P_2) := (P_1', P_2')$
   end while
   compute the new best bisection of G from (P_1, P_2)
end while
return (P_1, P_2)

When the entire procedure completes the best bisection found is returned. Function improve_and_cluster performs the following steps.

1. **Free all** vertices and set $c = 0$.
2. **Forward move**: Rank subsets V_1 and V_2 according to: (a) number of elements, (b) in case of a tie, according to the largest gain of a vertex in the subset, and (c) in case of a tie, randomly by tossing a balanced coin. Call the subset of larger rank F, and call the other subset T. Move a sequence of vertices $f_1, \ldots, f_k$ from F to T using a highest-gain-first scheme until either F is out of free vertices or a vertex with strictly positive gain is moved. Lock $f_1, \ldots, f_k$ in T, set $c = c + 1$, and let $L_c = \{f_1, \ldots, f_k\}$.
3. **Restore balance**: If $|V_1| \leq \alpha|V|$ and $|V_2| \leq \alpha|V|$ then do nothing. Otherwise, call F the larger of V_1 and V_2, and call the other subset T. Move a sequence of vertices $r_1, \ldots, r_j$ from F to T using highest-gain-first scheme until either F is out of free vertices or the current size of F is no more than $\alpha|V|$. Lock $r_1, \ldots, r_j$ in T, set $c = c + 1$, and let $L_c = \{r_1, \ldots, r_j\}$.
4. **Save**: If the current partition is an improved bisection then save it.
5. **Repeat**: If there are still free vertices then go to step 2.
6. **Clustering**: Cluster G by coalescing together some subsets of $G[L_1], G[L_2], \ldots, G[L_c]$.

If the best bisection is (P_1, P_2), found by improve_and_cluster, then subsets L_1, $L_2, \ldots$, and L_c are mutually disjoint—for each $1 \leq j \leq c$, either $L_j \subseteq P_1$ or $L_j \subseteq P_2$. And thus vertices that are coalesced together in step 6 of the procedure belong to the same subset in the best bisection (P_1, P_2); (P_1, P_2) clusters to a bisection (V_1', V_2') of the same cost in the clustered graph. Clustering is most effective when the size of each of the clustered subset is small in order to avoid entrapment in local minima, and when the number of clustered subset is large enough for efficient execution time. The above clustering approach achieves both of these goals.

In order to test the algorithm, the author has developed an algorithm with which he was able to generate input graphs for which optimal number of feedback arcs was known. A pseudo-code for this algorithm with which test cases were generated can be seen in the paper in question under section 7 (Generation of Test Cases). There were eighty random graphs generated, on 100, 500, 1000, and 2000 vertices, such that the average vertex out-degree was 2, 3, 4, 8, and 16. Divide-and-conquer algorithm was run using both SE and DC, so as to solve GB. Both strategies start with a random initial bisection, the algorithm was run 10 times. In each case, minimum, maximum, and mean of the size of the returned feedback arc set was calculated. The author has also implemented three other algorithms, and has made comparisons—these are ELS [47], ESL [103], and a modified version of ESL (MESL).

All the algorithms were implemented in the C programming language, and they were run on a Sun Sparc 1+ workstation. Comparing the algorithms by speed, the largest CPU time per run was at most 7 min. Linear time algorithm ELS was the winner, followed by ESL and MESL, which ran at about the same speed, these were followed by SE and DC with execution time dependent on the particular instance. The size of the smallest FAS found in 10 runs of DC and SE was never more than 8% and 11% above the optimal size for graphs, and was never more than 1.4% and 12.5% above the least known size. SE and DC typically performed better than ELS, ESL, and MESL, with DC being the more effective of the two approaches. Competing algorithms (ELS, ESL, and MESL) were performing the best on graphs with average degree 16 on graphs for which optimum was not known.

3.19 Fortran Subroutines for Computing Approximate Solutions of Feedback Set Problems Using GRASP [53]—December 2001

In this work the authors have proposed a Fortran subroutines for approximately solving the Feedback Vertex and Arc Set problems, on directed graphs, using a Greedy Randomized Adaptive Search Procedure (GRASP).[2] In the paper, the authors have also outlined implementation, as well as usage of the package. Computational experiments are reported, where solution quality is being illustrated as function of running time. Such NP-Hard problems are sometimes called Hitting Cycle problem, since one must hit every cycle in C.

GRASP is an iterative sampling method for finding approximate solutions to combinatorial optimization problems. Using GRASP, run is repeated, where each iteration is finding an approximate solution to the problem—therefore when speaking about GRASP we are speaking about a multi-start procedure. In terms of solution, the best solution found is what is returned by GRASP, this includes all iterations.

---

[2] On 26 August 2021, under a name 815.gz, a file for the subroutines was still available for downloaded at: https://calgo.acm.org/.

Each GRASP iteration is made up of two phases: a construction phase, and a local search phase (local improvement phase). During a construction phase, feasible solution is iteratively constructed. One element at a time is randomly chosen from a Restricted Candidate List (RCL), whose elements are ranked according to some greedy criterion, and is then added to the solution that is being built. Randomness of the procedure ensures variation in terms of the greedy choice, and thus the construction phase solution is rarely the greedy solution. After the first phase, the neighborhood of the constructed solution is searched for an improved solution, this step is the second phase of the method in question.

The authors of the paper have proposed gfvs and gfas, two sets of ANSI standard Fortran 77 subroutines that apply GRASP, in order to find approximate solutions of the Feedback Vertex and the Feedback Arc Set problems. In order to solve FAS the authors have used a helper procedure so as to translate one problem to another, and is such a way transform solutions. An $O(|E|)$ time procedure [50], is applied to translate the instance of FAS problem into an equivalent Feedback Vertex Set problem, which can then be solved by gfvs. And since there is a one-to-one correspondence between feasible solutions between FVS and FAS and their corresponding costs, an approximate solution to one problem can be translated into an approximate solution of the other problem. Procedure for problem transformation fas2fvs can be seen in Algorithm 3.19.1.

Algorithm 3.19.1 fas2fvs($G = (V, E), G' = (V', E')$)

$V' = E$
for all arc $e_i = (v_i, v_j) \in E$ **do**
   if $\exists\, e_j = (v_j, v_k) \in E$ **then**
      $E' = E' \cup \{(e_i, e_j)\}$
   end if
end for

For each arc in G there is a corresponding vertex in G', and for each pair of arcs in G, for which the head of the first arc is the tail of the second, there is an arc in G' whose tail is the vertex corresponding to the first arc in G and whose head is the vertex corresponding to the second arc.

The subroutines in files gfvs.f and gfas.f compute an approximate solution of the Feedback Vertex and Arc Set problems, respectively. Five parameters that control the execution of the algorithm need to be set before the optimization module is called: alpha (the restricted candidate list parameter: $\in [0, 1]$ or a negative value where each iteration uses a different randomly generated value), look4 (integer, stop if a cut-set of size at least look4 is found: $0 \leq look4 \leq |V|$ for FVS, or $0 \leq look4 \leq |E|$ for FAS), maxitr (integer, maximum number of iterations: $maxitr > 0$), prttyp (integer, output option parameter: 0 for silent run, or 1 for printing solution improvements, or 2 for printing found solution in each iteration), and seed (integer, pseudo-random number generator: $1 \leq seed \leq 2^{31} - 1$); default values for alpha, look4, maxitr, prttyp, and seed are: $-1, 0, 2048, 1$, and 270001; respectively.

By running feedback set procedures, on a subset of test problems used in [104], the authors have conducted their empirical experiment. The experiment was limited to half of the test problems, having at least 50 and at most 1000 vertices, with at most 30000 arcs, for a total of 40 instances. Both procedures were tested in order to ascertain performance and approximated solutions. Procedure gfas was tested on all instances of 50 and 100 nodes, but on instances which had 500 and 1000 nodes, and greater number of arcs, the procedure was not tested, and was in the experiment designated as DNR (Did Not Run).

Experiments were done on a Silicon Graphics Challenge computer with twenty 196 MHz MIPS R10000 processors, and 6.1 Gb of main memory. Code was compiled on the SGI Fortran compiler f 77 using the flags: -03 -r4 -64 -static. Processes were limited to a single processor, and CPU times were computed by calling the system routine etime(), computation was in seconds. Experiment has showed that while gfvs can be used on the entire set of test problems, gfas is only appropriate for solving sparse instances, since it requires long running times on dense problems. Procedure gfas creates an equivalent Feedback Vertex Set problem with as many nodes as the number of arcs of the graph in the original problem, and this FVS problem is then solved by GRASP for FVS. Therefore it was expected that a subroutine for FAS would exhibit such a behavior.

3.20 A New Rounding Procedure for the Assignment Problem with Applications to Dense Graph Arrangement Problems [9]—March 2002

In this work one can find a randomized procedure for rounding fractional perfect matchings (includes every vertex) to integral matchings. Developed rounding procedure was used to design an additive approximation algorithm in order to solve Quadratic Assignment problem (QA). Rounding procedure for small coefficients consists of the following phases: decomposition and merge. The more general procedure where the value of the coefficients is allowed to grow with n consists of the same two phases with a modification of a merge phase, that is merge operator. In the former phase merge operator on matchings took the union of two matchings, and broke long cycles/paths at arbitrarily chosen points, and so the resulting paths/cycles are all "small." The latter merge phase chooses breakpoints randomly.

Rounding procedure presented can be implemented efficiently on a parallel multiprocessor, i.e., it can be made to run on an EREW PRAM in $O(\log n \log \log n)$— therefore in RNC^2 complexity class.

When dealing with assignment problem, one wishes to minimize a linear cost function over the set of permutations. In the paper in question a variant is considered in which the perfect matching has to satisfy more than one linear constraint—the Assignment Problem with Extra Constraints (APEC).

Approximation error of the designed algorithm is ϵn^2 with algorithm complexity being $n^{O(\log n/\epsilon^2)}$, where n represents number of nodes. An addition to the aforementioned is Quasi-Polynomial Time Approximation Schemes for "dense" instances of a number of well known NP-Hard arrangement problems: minimum linear arrangement, min-cut linear arrangement, maximum acyclic subgraph, betweenness, graph isomorphism, d-dimensional arrangement.

Denseness is defined as per problem in question, if one deals with a graph, then it means that the average degree is $\Omega(n)$, and that the number of edges is at least $a \cdot n^2$ for a graph that is a-dense. Approximation schemes presented by authors compute a $(1 + \epsilon)$-approximation in running time mentioned above. For some problems one can reduce a space on which exhaustive search is conducted, and in such cases authors give algorithms that run in $n^{O(1/\epsilon^2)}$ time, these are therefore Polynomial Time Approximation Schemes (PTAS). Algorithms presented are randomized, but can be derandomized using techniques in [60, 95], whether betweenness could be derandomized was not answered.

PTAS algorithms described in the paper, which includes the one we are interested in, namely Maximum Acyclic Subgraph problem (a dual of FAS), are randomized, and as such they are, as envisioned, choosing a random multi-set of $O(\log n)$ vertices to work on. As per transformation requirement, FAS firstly needs to be rephrased. Pseudo-code for the algorithm is not given in the paper in question, and so as not to repeat the text of the procedure we are directing a reader to the paper in question, please look under subsection 4.3 with the title Maximum Acyclic Subgraph.

3.21 On Enumerating All Minimal Solutions of Feedback Problems [119]—March 2002

Here, an algorithm that generates all minimal Feedback Vertex Sets of a directed graph $G = (V, E)$ is presented. These sets are generated with a polynomial delay $O(|V|^2(|V|+|E|))$. Underlying technique for solving FVS can be bespoken in order to generate all minimal solutions for both undirected and directed Feedback Arc Set—this can be achieved with polynomial delay $O(|V||E|(|V|+|E|))$. Computing the number of minimal feedback arc sets of a directed graph is #P-Hard.

The algorithm for minimal FVS relies on an exhaustive search in a superstructure graph Φ, whose vertices represent minimal feedback vertex sets of G. This Φ is strongly connected and has a diameter of at most $|V|$. The algorithm for enumerating minimal feedback arc sets can also be applied for enumeration of minimal transversals of directed planar graphs $G = (V, E)$—where a transversal is a set of arcs that contains at least one arc of each directed cut. Since a directed cut in a planar graph G corresponds to the set of arcs in a directed cycle in its dual graph H, a transversal in G corresponds to a feedback arc set in H and vice versa.

In order to compute all minimal feedback arc sets of G a function for generating minimal feedback vertex sets can be used. This function can generate all minimal solutions by an exhaustive search in the superstructure graph $\Phi(G, \mu_G)$. The vertex set of $\Phi(G, \mu_G)$ consists of all MFVSs F of G. For each F there are directed arcs from F to each μ_G-successor of F. If one starts with an initial MFVS $F = F_0$, all successors of F are generated, that is F is being expanded, and after this step, solutions that are still unexpanded are determined—this process is repeated until all generated solutions have been expanded. An algorithm for MFVS generation is presented in Algorithm 3.21.1.

Algorithm 3.21.1 GENERATE-MFVS(G, μ_G)

---

compute a minimal admissible solution F_0
insert F_0 into Q and into D {algorithm uses a queue Q and a dictionary D}
while Q is not empty **do**
    remove any set F from Q
    output F
    for all μ_G-successor F' of F **do**
        if F' is not contained in D **then**
            insert F' into D and Q
        end if
    end for
end while

When one starts from a given feedback vertex set X, a minimal FVS $F' \subseteq X$ can be computed by verifying for each $v \in X$ whether $X \setminus v$ is FVS for G. If this is true, then v is removed from X. After this operation has been completed for each $v \in X$, once, the remaining feedback vertex set $F' \subseteq X$ is minimal. A single operation for $v \in F$ can be performed using DFS in time $O(|V|+|E|)$, thus the FVS minimization can be performed in $O(|V|(|V| + |E|))$—which makes running time of one execution of while loop in Algorithm 3.21.1 stand as $O(|V|^2(|V|+|E|))$, this execution time makes aforementioned polynomial delay for the successive output of minimal FVSs. Memory requirements are polynomial for graphs with a polynomial number of minimal FVS, but potentially exponential for the general case.

Feedback arc sets of a graph and feedback vertex sets of its line graph have a close relationship. The line graph G' of a digraph $G = (V, E)$ is a digraph G' that has a vertex $v'(e)$ for each arc $e \in E$, and an arc $e' = (v'(e_1), v'(e_2))$ for any two arcs $e_1 = (x, y) \in E$ and $e_2 = (y, z) \in E$. Each cycle in G corresponds to a cycle in G', and vice versa. Therefore, feedback arc sets of G correspond to feedback vertex sets of G'—considering that G' has $O(|E|)$ vertices and $O(|E|^2)$ arcs, FASs of G can be calculated with a time complexity of $O(|E|^4)$ per minimal feedback arc set. The procedure for FAS, a variation of the aforementioned, is very similar to procedure for FVS, vertices and arcs are swapped, and uses time $O(|V||E|(|V| + |E|))$ per minimal solution. The procedure that transforms MFAS F into a F^* by generating μ_G-successors can be seen in Algorithm 3.21.2.

Algorithm 3.21.2 TRANSFORM-DIRECTED-MFAS($G = (V, E), F, F^*, \mu_G$)

---

compute a topological order $\mathcal{T}$ of $G \setminus F^*$
$F_0 := F, k := 0$
while $F_k \neq F^*$ **do**
    let v_k be the minimal vertex of $S(F_k \cap (E \setminus F^*))$ with respect to $\mathcal{T}$
    $F_{k+1} := \mu_G(F_k, v_k)$
    $k := k + 1$
end while
output $(F_0, \ldots, F_k)$

For insight into how the acyclic orientations of undirected graph $G = (V, E)$ correspond to the minimal feedback arc sets of a closely related directed graph $\bar{G}$ one can consult section 5 of the paper in question, with the title Feedback Arc Sets and Acyclic Orientations. For a related work, and a note on MFAS please consult [40].

3.22 Combinatorial Algorithms for Feedback Problems in Directed Graphs [39]—May 2003

The authors of the paper have presented simple combinatorial approximation algorithms for FAS and FVS that are obtaining solutions with an approximation ratio bounded by the length λ of a longest simple cycle of the digraph—in terms of number of arcs. Approaches that the authors have developed are built on top of the local-ratio technique found in [13] (other main approach found in the literature is the primal-dual [13]). They have a complexity of $O(m \cdot n)$ on a digraph with n vertices and m arcs. In preliminary experimental research, in a crossing minimization application, the algorithms have proved to be practical on dense instances with many short cycles.

If one is dealing with covering problems, then Local-Ratio theorem can be informally stated as follows. If a cover C is a r-approximation with respect to both weight functions w_1 and w_2, then C is r-approximation with respect to the weight function $w_1 + w_2$. That is, if the payment at each step can be proved to cost no more than r times the optimum payment, then the total payment will be at most r times the optimum cost.

Foundational algorithm is the algorithm for FAS. This algorithm can then be adapted to serve FVS as well. General approach consists of progressively reducing the weights of the arcs of the digraph, and then adding to the feedback arc set those arcs whose weight became equal to 0. The algorithm consists of two phases, in the first phase the algorithm is trying to acquire FAS and make digraph acyclic. In the second phase the algorithm is trying to introduce some arcs back into digraph in order to make FAS minimal if possible, while at the same time avoiding making digraph cyclic again. The set of removed arcs is returned in the end. Pseudo-code for the procedure can be seen in Algorithm 3.22.1.

Algorithm 3.22.1 MFAS of a weighted digraph—$FAS(G = (V, A), w : A \rightarrow \mathbb{R}^+)$

$F \leftarrow \emptyset$ {F is the feedback arc set found by the algorithm}
Phase 1
while $((V, A \setminus F)$ is not acyclic) **do**
   let C be a simple cycle in $(V, A \setminus F)$
   let (x, y) be a minimum weight arc in C and let ϵ be its weight
   for all $(v, w) \in C$ **do**
      $w(v, w) \leftarrow w(v, w) - \epsilon$
      if $w(v, w) = 0$ **then**
         $F \leftarrow F \cup \{(v, w)\}$
      end if
   end for
end while
Phase 2
for all $(v, w) \in F$ **do**
   if $(V, A \setminus F \cup \{(v, w)\})$ is acyclic **then**
      $F \leftarrow F \setminus \{(v, w)\}$
   end if
end for
return F

The algorithm is trying to find a compromise between two sub-objectives, removing light arcs (arcs with small weight), and removing arcs belonging to a large number of cycles. Algorithm is working on decreasing weight of all the arcs in any cycle it finds. The bigger is the number of cycles an arc belongs to, the more likely is the reduction of its weight, and the more likely is its subsequent removal. The algorithm first finds an optimal solution to a relaxed integer programming formulation of FAS, and uses it to partition the set of vertices into two disjoint sets V_1 and V_2. Then, it deletes the cheapest set of arcs either from V_1 to V_2 or vice versa—after that, it recourses both on V_1 and on V_2. To improve algorithm performance one could, in phase 1, use a heuristic for choosing the shortest available cycle, or to potentially improve solution quality, one could use a heuristic for ordering arcs by decreasing weight, in phase 2.

3.23 A Contraction Algorithm for Finding Minimal Feedback Sets [86]—January 2005

The author of the paper has presented a contraction algorithm for finding a minimal set of arcs/vertices (for reduction between FAS and FVS one can also consult [50, 52]) that break all cycles in a digraph—this algorithm represents an extension of a contraction algorithm for FVS published in [94].

A modified version of the original contraction algorithm offers better worst-case performance when used for FAS, and it also preserves reducibility property. It does not, however, extend the class of graphs for which it finds a minimal feedback set, although with some additional steps the algorithm can be extended in order to allow for both feedback arcs and vertices. The algorithm is also working on Reducible Flow Graph, since it preserves the structure, and is easily combined with the algorithm published in [110].

The algorithm is designed to work with and without weights, and has time complexity of $O(m \log n)$, where m is a number of arcs and n represents number of vertices. For a case of $FAS \rightarrow FVS$, which enlarges the size a graph, the overall complexity for an unweighted FAS-graph is $O(nm \log m)$.

A series of steps for the FVS contraction algorithm can be seen in section 3 of the paper in question, Contraction Algorithms for the FVS Problem. Steps for the extended algorithm, which can be implemented to run in $O(m \log n)$, that allows for both arcs and vertices, are as follows:

1. **loop-arc**(e): if $e : (v_\infty \rightarrow v_\infty)$ is a loop and v_∞ an ∞-vertex then remove e from the graph and add it to the feedback set
2. **loop**(v): if v contains a loop of infinite weight then remove v from the graph and add it to the feedback set
3. **in0**(v): if v has no incoming arcs then remove v from the graph
4. **out0**(v): if v has no outgoing arcs then remove v from the graph
5. **in1-fin**(v): if v has only one incoming arc e and $weight(e) < weight(v)$ then subst(v, e)
6. **out1-fin**(v): if v has only one outgoing arc e and $weight(e) < weight(v)$ then subst(v, e)
7. **in1-inf**(v): if v has only one incoming arc $e : (v' \rightarrow v)$, $v \neq v'$, e has infinite weight and $weight(v') \leq weight(v)$ then remove v and e from the graph and move all outgoing arcs of v to become outgoing arcs of v'
8. **out1-inf**(v): if v has only one outgoing arc $e : (v \rightarrow v')$, $v \neq v'$, e has infinite weight and $weight(v') \leq weight(v)$ then remove v and e from the graph and move all incoming arcs of v to become incoming arcs of v'
9. ∞-**cycle**(e): if $e : (v \rightarrow v_\infty)$ has an inverse arc $e' : (v_\infty \rightarrow v)$ and e, e' and v_∞ have infinite weight then remove v from the graph and add it to the feedback set
10. ∞-**mark**(e): if $e : (v_1 \rightarrow v_2)$ has finite weight such that $weight(e) \geq weight(v_1)$ or $weight(e) \geq weight(v_2)$ then subst(e, ∞)
11. **io1**(v): if v has only one incoming arc $(v_1 \rightarrow v)$ and one outgoing arc $(v \rightarrow v_2)$, both of infinite weight, then remove v from the graph and add an arc $e : (v_1 \rightarrow v_2)$ instead, and subst(e, v)
12. **parallel**(e_1, e_2): if $e_1 : (v_1 \rightarrow v_2)$ and $e_2 : (v_1 \rightarrow v_2)$ are parallel arcs then subst$(e_1, \{e_1, e_2\})$ and remove e_2 from the graph.

Algorithm, and its variations, have been implemented and tested on randomly generated RFGs. RFGs were of different sizes and densities, and where both weighted and unweighted, with unweighted having all arcs of the same wight.

Experiment was conducted on 12 test cases with each test case having 1000 randomly generated graphs. Algorithm for random graph generation can be found in section 7, Experimental Results. The extended algorithm version performed quite well for unweighted RFGs, although its performance for the weighted case was not on the same level (test was performed on a 1.8 GHz PC).

3.24   Improved Exact Exponential Algorithms for Vertex Bipartization and Other Problems [114]—October 2005

The authors of this paper have presented a number of exact algorithms for NP-Hard problems. These are an $O(1.9526^n)$ algorithm for Vertex Bipartization problem in undirected graphs, an $O(1.8384^n)$ algorithm for Vertex Bipartization problem in undirected graphs with maximum degree of 3, an $O(1.945^n)$ algorithm for Feedback Vertex Set and Vertex Bipartization problem in undirected graphs with maximum degree of 4, an $O(1.9799^n)$ algorithm for 4-Hitting Set problem, and an algorithm, that is of interest to us, with a time complexity of $O(1.5541^m)$ for minimum Feedback Arc Set problem in tournaments—take note that in all these algorithm complexities polynomial terms are suppressed, and that n represents a number of vertices, and m is a number of edges. In order to obtain an exact algorithm the authors present two general techniques, one is parameterized complexity, and the other is "colored" Branch and Bound.

Exact algorithms for hard to solve problems are generally too inefficient, nevertheless for problems that are small to moderate in size these algorithm can have practical value. Therefore it is worthy to try to make exact algorithms as efficient as possible, and in this way provide an avenue through which certain practical instances can be solved optimally.

For a theoretical background for the FAS algorithm one can look in theorem 4 and section 3.1, Applications, subsection Feedback Set Problems in Tournaments, of the paper in question. For a related algorithm previously published one is refereed to [112]. General exact procedure can be seen in Algorithm 3.24.1. It should be noted that the authors have not investigated practical performance of the algorithms presented.

In [115] and extended version of the paper can be found, the paper is titled, Efficient Exact Algorithms Through Enumerating Maximal Independent Sets and Other Techniques. In this extended version a new technique for algorithm design has been expounded, namely enumeration of maximal independent sets in a graph. The authors have also devised new algorithms for a few other problems as well. A survey of the results of the extended paper can be seen in Table 3.1.

Algorithm 3.24.1 Exact(Q, $\mathcal{A}$, c)

Input: Q is a minimization problem and $\mathcal{A}$ is the Fixed-Parameter Tractable (FPT) algorithm that solves its parameterized version in time $O^*(c^k)$, where c is a constant and k is the parameter—n is the size of the universe U.
Output: Compute te largest λ such that $c^{\lfloor n\lambda \rfloor} \leq \binom{n}{n - \lfloor \lambda n \rfloor}$.

for $i = 1$ **to** $\lfloor \lambda n \rfloor$ **do**
    use the FPT algorithm $\mathcal{A}$ for Q to check whether there is solution of size i; if yes, then output
    i and halt.
end for
for $i = \lfloor \lambda n \rfloor + 1$ **to** n **do**
    try all subsets of size i of U to check whether there exists a solution of size i; if yes, then
    output i and halt.
end for

Table 3.1 A survey of algorithms and results thereof

Problem	Previous best	Authors result
ODD CYCLE TRANSVERSAL (undirected graph)	$O^*(1.7724^n)$	$O^*(1.9526^n)$
ODD CYCLE TRANSVERSAL (undirected graph; minimum sized)		$O^*(1.62^n)$
ODD CYCLE TRANSVERSAL (undirected graph; count all minimum sized)	$O^*(1.7724^n)$	$O^*(1.6713^n)$
FEEDBACK VERTEX SET (digraph with total degree ≤ 4)	$O^*(2^n)$	$O^*(1.945^n)$
FEEDBACK ARC SET (tournament)	$O^*(2^m)$	$O^*(1.5541^m)$
MINIMUM MAXIMAL MATCHING (undirected graph)	$O^*(1.4422^m)$	$O^*(1.44225^n)$
MINIMUM EDGE DOMINATING SET	$O^*(1.4422^m)$	$O^*(1.44225^n)$
4-HITTING SET	$O^*(2^n)$	$O^*(1.9646^n)$

3.25 Parameterized Algorithms for Feedback Set Problems and Their Duals in Tournaments [113]—February 2006

FVS and FAS on a well known class of digraphs, namely tournaments, is studied—where both weighted and unweighted versions are considered. Aside from these problems the authors have also studied their duals, that is, given a digraph a question is raised: Is there a subset of k vertices/arcs that forms a digraph which is acyclic and is a subgraph of the graph?

Among main scientific contributions of the paper are an algorithm for weighted feedback arc set problem in tournaments whose complexity is $O(2.415^k n^w)$ where w is the exponent of the running time of the best matrix multiplication algorithm, and an algorithm for the dual of FAS (maximum, arc induced, acyclic subgraph) in general directed graphs with time complexity of $O(4^k k + m)$. It has been additionally shown that the problem of FAS is fixed-parameter tractable in dense directed graphs which are at most $n^{1+o(1)}$ arcs away from a tournament. Algorithm for FAS which is improved and has better branching technique is presented in Algorithm 3.25.1.

Algorithm 3.25.1 BTFAS(T, k, F)

Input: T is a tournament, $k \geq 0$, F is a set of arcs. Initially the algorithm is called by $BTFAS(T, k, \emptyset)$.

Output: Returns 'true' and a minimal feedback arc set of size at most k if one exists, returns 'no' otherwise. F contains the arcs of a partial feedback arc set that are reversed from the original T.

S0: If T does not have a directed triangle, then return 'true' and F

S1: If $k = 0$ and T has a triangle, then answer 'no' and 'exit'

S2: Find an induced subgraph on 4 vertices in T, isomorphic to F_1, if one exists. Such a subgraph is simply a tournament on 4 vertices having at least two directed triangles. Let the vertex set of such an F_1 be $\{1, 2, 3, 4\}$—in particular $(1, 2)$ is the only arc not part of any directed triangle. If no such subgraph exists in T, then go to S6.

S3: Let $\{a, b, c\}$ be the arcs of a triangle in F_1, such that there exists an arc $x \in \{a, b, c\}$ for which $rev(x) \in F$. If there is no such triangle in F_1, then go to S4.

  S3a: If $rev(a)$, $rev(b)$ and $rev(c)$ are in F, then answer 'no' and 'exit'

  S3b: If $BTFAS(T \setminus \{x\} \cup rev(x), k - 1, F \cup \{x\})$ is true for any arc x of the triangle such that $rev(x)$ is not in F, then return 'true' and F (here F is $F \cup \{x\}$) and 'exit', else answer 'no' and 'exit'

S4: If $rev((1, 2)) \notin F$ and if any of the following recursive calls returns 'true', then return 'true' and the corresponding F and 'exit', and answer 'no' and 'exit' otherwise $\{T'$ is obtained from T by reversing the 'newly included' arcs of F. $BTFAS(T', k - 1, F \cup \{(3, 4)\})$, $BTFAS(T', k-2, F\cup\{(4, 1), (4, 2)\})$, $BTFAS(T', k-2, F\cup\{(4, 1), (2, 3)\})$, $BTFAS(T', k-2, F \cup \{(1, 3), (2, 3)\})$, $BTFAS(T', k - 3, F \cup \{(1, 2), (1, 3), (4, 2)\})$.$\}$

S5: If $rev((1, 2)) \in F$ and if any of the first 4 recursive calls enumerated in S4 returns 'true', then return 'true' and the corresponding F and 'exit', otherwise answer 'no' and 'exit'

S6: Find a minimum FAS S of a resultant tournament in polynomial time. If a tournament does not have directed cycle of length 4, then no pair of directed triangles in the tournament has a vertex in common. Hence the minimum weight feedback vertex or arc set is obtained by finding all triangles, and picking a minimum weight vertex/arc from each of them. This can be done by computing the square of the adjacency matrix of the tournament. If $|S| > k$ then return 'no' and 'exit', else return 'true' and $F \cup S$ and 'exit'.

Tournament T has a directed cycle of length 4 iff it has a subgraph isomorphic to F_1—see Fig. 3.1—and F_1 can be found in $O(n^w)$ time. Algorithm 3.25.1 can also be applied for weighted FAS problem in a tournament where the weight of every arc is at least 1.

Fig. 3.1 A directed graph F_1

It is perhaps noteworthy to mention, where a close link between feedback arc set and feedback vertex set plays a significant role, that the authors have also given FPT algorithm for the parametric dual of directed feedback vertex set in oriented directed graphs where cycles of length 2 are not allowed. An oriented directed graph is a directed graph where there is at most one directed arc between every pair of vertices.

3.26   A Branch-and-Bound Algorithm to Solve the Linear Ordering Problem for Weighted Tournaments [28]—October 2006

In this paper the authors have presented principles and results of an exact method designed in order to solve the Linear Ordering problem for any weighted tournament T. The algorithm devised is based upon a Branch and Bound search with a Lagrangean relaxation used for evaluation, and a noising method used for computation of initial bound. Along with these the authors have also implemented other components so as to reduce the BB-search tree.[3]

For small enough instances the algorithm is almost always finding solutions that are optimal, and this is done in a short time. The authors expect this empirical result to translate for larger instances as well. In order to use Lagrangean relaxation and the noising method one has to tune several parameters. These parameters are functions of features of a tournament to be solved: number of vertices, weights, and index of transitivity. Therefore in order to run the method one has to only input number of vertices, the orientations and the weights of the arcs of a tournament. It is, however, possible to change parameter values so as to improve method results. For example, the noising part of the algorithm is benefiting with number of iterations increase, since giving more time (more iterations) increases probability of getting an optimal solution.

Lagrangean relaxation part of the Branch and Bound algorithm is based upon a work done by Arditti [7] who has studied application of it on the Linear Ordering problem. Arditti used it as a heuristic in his approach, and not as a part of a Branch and Bound type of algorithm. The authors of the paper in question used it in an improved and generalized way so as to evaluate nodes in the search tree of the Branch and Bound process. In order to estimate a first bound, that is approximate solution, a noising method is applied—for more details on this metaheuristic one can consult [27]. A general goal of the procedure is to iteratively and locally transform solution after solution, and in such a way achieve intended objective after a certain number of steps. Into this process a noise is added, according to a certain probability

---

[3] The software was at the time of publication of the paper in question available at: http://www.enst.fr/~charon/tournament/median.html. It seems, however, that it is not available anymore, access was attempted on 27th of September 2021.

distribution and with regards to a certain interval. In this way it is possible to reject a good solution, as well as accept a bad one. To tackle this issue noise is gradually diminished, until it reaches 0 and only a pure solution is being dealt with.

Empirical analysis of the algorithm has been conducted. The experiments were done on a SUN Sparc workstation. Different components of the algorithm were tested on three graphs, but the claim is that other tests lead to the same qualitative results. CPU times needed to find a solution for these graphs were 509 s, 2208 s and 2451 s. In order to ascertain efficiency of the algorithm components authors have computed number of nodes cut by each component at each depth in Branch and Bound search tree—there are six components in total through which the algorithm is trying to cut a branch. It needs to be noted that the results can be quite different if one changes a sequence in which these components are applied—for more details please consult figures 2–4 of the paper in question.

Developed software was also tested on 5790 tournaments with different charac-
teristics. For each set of characteristics 30 random tournaments were generated, with these characteristics. For each instance CPU time was recorder, which represented time needed to solve this instance, and then an expected value was calculated—there were in total 193 sets of tournaments. Depending on the type of tournament T and a number of vertices n, average CPU time measured in seconds, required to solve problem instances ranged from thousands of seconds for more difficult problems, to hundreds of seconds for those instances that were less difficult—with difficulty depending on transitivity index $\tau(T)$ calculated by taking into account number of 3-circuits of T, the lower transitivity index is, the more difficult instance is to solve. In real-life instance this index is high and often very close to 1, therefore it is often possible to solve these real-life problems with high orders in mind.

$$\tau(T) = 1 - \frac{\text{number of 3-circuits of } T}{\text{maximum number of 3-circuits of a tournament of order } n} \tag{3.5}$$

As for the noising component of the algorithm, which is run before branching starts, over the 5790 tournaments tested and with up to 100 vertices, the noising method found optimal solution in almost all cases, all but six to be exact—where second application of the noising method succeeded in solving these 6 tournaments exactly. In this way noising method gave optimal solution for approximately 99.9% of the studied problems, for the orders considered in the experiment. The method is efficient, and it never takes more than a few seconds to execute, it is also stochastic in nature and can therefore, if necessary, be rune multiple times, and in this way amplify probability to obtain quality solution. For related research please consult [106, 107]

3.27   How to Rank with Fewer Errors—A PTAS for Feedback Arc Set in Tournaments [83, 97]—June 2007

Authors of this work have presented Polynomial Time Approximation Scheme for the minimum Feedback Arc Set problem in tournaments. This is the first FAS work of such kind. The authors have also devised a weighted generalization which also represents first PTAS for Kemeny Rank Aggregation (KRA).

As the authors have described so simply and eloquently what Feedback Arc Set on tournament really is, we will give this part of the authors text in its entirety, for some general tournament.

> Suppose you ran a tournament, everybody played everybody, and you wanted to use the results to rank everybody. Unless you were really lucky, the results would not be acyclic, so you could not just sort the players by who beat whom.
>
> A natural objective is to find a ranking that minimizes the number of upsets, where an upset is a pair of players where the player ranked lower in the ranking beat the player ranked higher. This is the minimum Feedback Arc Set (FAS) problem in tournament.

The main theorem of the paper states that there is a randomized algorithm for minimum Feedback Arc Set on weighted tournaments. Given $\epsilon > 0$, it outputs a ranking with expected cost at most $(1 + \epsilon)OPT$. The expected running time of this algorithm is $O(n^3 \log n(\log(1/b) + 1/\epsilon)) + n(2^{\tilde{O}(1/(\epsilon b)^6)})$. This algorithm can be derandomized at the expense of increasing algorithm complexity to $n^{\tilde{O}(1/(\epsilon b)^{12})}$. This PTAS is singly exponential in $1/\epsilon$, as opposed to a PTAS in the conference version of this work which was doubly exponential in $1/\epsilon$ [83].

As a consequence of the previous, there is a randomized algorithm for KRA that, given $\epsilon > 0$, outputs a ranking with expected cost at most $(1 + \epsilon)OPT$. Expected running time for n candidates is $O\left(\frac{n^3 \log n}{\epsilon}\right) + n2^{\tilde{O}(1/\epsilon^6)} + O(n^2 \cdot (number\ of\ voters))$. This algorithm can as well be derandomized at the cost of increasing complexity of the algorithm to $n^{\tilde{O}(1/\epsilon^{12})}$.

In order to solve the problem with whom they have taken an issue with, the authors have used three, as they called it, algorithmic tools: local search, KwikSort algorithm [3] and sampling-based approximation algorithm [98]. In a local search, a single vertex move, for a ranking π, a vertex x and a position i, consists of taking x out of π and putting it back in position i. A ranking is then locally optimal if no other single vertex move can improve the cost of the ranking. The main algorithm can be seen in Algorithm 3.27.1, the KwikSort algorithm for MFAS on tournaments can be seen in Algorithm 3.27.2.

The algorithm for bucketed FAS with additive error $O(\delta n^2)$ can be seen in Algorithm 3.27.3. This algorithm is a specialization of the algorithm in [98] for

---

Algorithm 3.27.1 Approximation scheme for minimum FAS on a tournament

---

run the *KwikSort* algorithm on V to define ranking π
return *Improve*$(\pi, eb/5)$

procedure *Improve(ranking π, error tolerance η)*
set $\beta \leftarrow \frac{\eta C(\pi)}{4n \log_{3/2} n}$ and $\kappa \leftarrow \frac{n^2 b^3}{350 \cdot 400^2}$
perform single vertex moves on π until none can improve the cost by more than β
return *ImproveRec*(V, π)
end *Improve*

procedure *ImproveRec(vertices S, ranking π on S)*
if $|S| = 1$ **then**
  return π
else
  if $C(\pi) \geq \kappa |S|^2$ **then**
    return the ranking from *AddApprox* with $\delta = \frac{\eta}{4}\kappa$
  else
    choose an integer k uniformly at random from $[|S|/3, 2|S|/3]$
    let L be the set of vertices v such that $\pi(v) \leq k$ and $R = S \setminus L$
    return concatenation of *ImproveRec*(L, π_L) and *ImproveRec*(R, π_R), where π_L is the
    ranking of L induced by π (i.e. $\pi_L(v) = \pi(v)$), and π_R is the ranking of R induced by π
    (i.e. $\pi_R(v) = \pi(v) - k$)
  end if
end if
end *ImproveRec*

Algorithm 3.27.2 KwikSort(*vertices S*)

---

choose a vertex v uniformly at random from S
let L be the set of vertices u such that $w_{uv} \geq w_{vu}$ and $R = S \setminus L$
return the concatenation of *KwikSort*(L) and *KwikSort*(R)

---

the bucketed FAS. The difference is in $|T| = O(1/\delta^2)$ [98], while here $|T| = \tilde{O}(1/\delta^2)$—in a specialized version there are number of buckets that depend on δ, while in [98] an assumption is made that the variables have constant-sized domains.

The procedure Algorithm 3.27.1 (FAST-Scheme) can be made more efficient. Instead of calling Improve with error tolerance $\epsilon b/5$, it can first be called $\log(1/b)$ times with error tolerance of $1/2$, before running it once with error tolerance of $\epsilon/7$. Secondly, in order to bound a total running time of single vertex moves the cost needs to be monotone non-increasing. For the purpose of achieving this, Improve is modified by replacing the line "**return** ImproveRec(V, π)" with the line "**return** either π or ImproveRec(V, π)", depending on cost that is lower. This faster scheme is a $1 + \epsilon$-approximation and can be seen in Algorithm 3.27.4. Algorithm complexity of this scheme is stated above, and for clarity is repeated here, $O(n^3 \log n (\log(1/b) + 1/\epsilon)) + n(2^{\tilde{O}(1/(\epsilon b)^6)})$.

In order to derandomize FAST-Scheme the authors have made three changes: AddApprox was replaced with deterministic additive error algorithm found in [57], KwikSort (which was derandomized in [136, 137]) was replaced with the

---

Algorithm 3.27.3 AddApprox—bucketed FAS with additive error $O(\delta n^2)$

---

take a sample S of $\tilde{O}(1/\delta^4)$ vertices chosen uniformly at random without replacement from V
take a sample T of $\tilde{O}(1/\delta^2)$ vertices chosen uniformly at random without replacement from S
for all of the possible bucketed rankings of T **do**
    for all vertex v of $S \setminus T$ in random order **do**
        place v in the bucket that minimizes the cost (of the bucketed ranking of the vertices placed
        so far)
    end for
end for
extend the discovered bucketed ranking of S with the minimum cost as follows
for all vertex v of $V \setminus S$ in random order **do**
    place v in the bucket that minimizes the cost so far
end for

Algorithm 3.27.4 FASTer-Scheme—faster approximation scheme with error tolerance $\epsilon > 0$

---

run $KwikSort$ to define a ranking π
for $i \leftarrow 1$ to $\lceil \log_2 1/b \rceil$ **do**
    $\pi \leftarrow Improve(\pi, 1/2)$
end for
return $Improve(\pi, \epsilon/7)$

---

deterministic constant-factor approximation algorithm, and randomized choice of k in ImproveRec was eliminated. In order to derandomize ImproveRec, every k was tried, and the best ranking found was kept. To prevent exponential expansion during run-time intermediate results were cached, and in this way a method known as Dynamic Programming was applied. This derandomized version can be seen in Algorithm 3.27.5. Excluding cache loops ImproveRec runs in $O(n^2)$, where each call has run-time dominated by the derandomized additive error algorithm with running time $n^{\tilde{O}(1/((\epsilon b)^3)^4)} = n^{\tilde{O}(1/(\epsilon b)^{12})}$. The overall complexity is therefore $O(n^2) \cdot n^{\tilde{O}(1/(\epsilon b)^{12})} \rightarrow n^{\tilde{O}(1/(\epsilon b)^{12})}$. As far as the run-time of the derandomized FAST-Scheme is a concern, it is dominated by $O(\log 1/b)$ calls to procedure Improve, and so the overall running time is $(\log 1/b)n^{\tilde{O}(1/(\epsilon b)^{12})} \rightarrow n^{\tilde{O}(1/(\epsilon b)^{12})}$.

3.28   Feedback Arc Set Problem in Bipartite Tournaments [64]—February 2008

The author of this paper has presented ratio-4 deterministic and randomized approximation algorithms for the FAS problem in bipartite or multipartite tournaments. A directed graph $D = (V, A)$ is a multipartite tournament if D is a directed orientation of a complete k-partite graph. If the situation is that $k = 1$ then it is an orientation of complete undirected graph and is known as a tournament—naturally for $k = 2$ it is called a bipartite tournament.

Algorithm 3.27.5 ImproveRec($vertices$ $S, ranking$ π on S)—derandomized version

---

if $ImproveRec(S, \pi)$ previously computed **then**
    return cached π^{out}
else if $|S| = 1$ **then**
    return π
else if $C(\pi) \geq \kappa |S|^2$ **then**
    return ranking from the deterministic additive error algorithm [57] with $\delta = \frac{\eta}{4} \cdot \kappa$
else
    initialize π^{out} to an arbitrary ranking (e.g. π)
    for $k = \lceil |S|/3 \rceil$ **to** $\lfloor 2|S|/3 \rfloor$ **do**
        let $L = \{v \in S : \pi(v) \leq k\}$ and $R = S \setminus L$
        let π_L be the ranking of L induced by π (i.e. $\pi_L(v) = \pi(v)$)
        let π_R be the ranking of R induced by π (i.e. $\pi_R(v) = \pi(v) - k$)
        let π^{temp} be the concat. of $ImproveRec(L, \pi_L)$ and $ImproveRec(R, \pi_R)$
        if $C(\pi^{temp}) < C(\pi^{out})$ **then**
            $\pi^{out} \leftarrow \pi^{temp}$
        end if
    end for
    return π^{out}
end if

The problem of FAS can also be stated as an ordering problem, where the arcs from higher indices to lower indices, called backward arcs, are minimized. Directed graph D has a topological ordering iff D is a directed acyclic graph. This equivalent formulation of FAS is the one that the author has used and found useful for devised algorithms, which will always output an ordering of vertices as solution—where the feedback arc set F can be formed by taking all the backward arcs in this ordering.

Randomized approximation algorithm for FAS uses the fact that a bipartite tournament has directed cycles iff it has a directed 4-cycle. The algorithm works in such a way where it randomly selects an arc, and partitions the bipartite tournament into two smaller ones, around this arc, and then solves the problem on these two smaller bipartite tournaments in a recursive manner.

In any digraph the size of minimum feedback arc set is at least the number of maximum arc disjoint directed cycles. Considering that maximum number of arc disjoint 4-cycles is upper bounded by the size of maximum arc disjoint cycles one has the following formula, $|MAD4C| \leq |MADC| \leq |MFAS|$. This $|MAD4C|$ is used as a lower bound for the algorithms, and the author has showed that the feedback arc set outputted by the algorithms is at most 4 times the size of $MAD4C$. This randomized approximation algorithm for FAS can be seen in Algorithm 3.28.1.

Along with randomized algorithm for Feedback Arc Set problem in bipartite tournaments, the author has also presented a deterministic factor-4 approximation algorithm for Feedback Arc Set problem in bipartite tournaments. This algorithm is basically a derandomized version of the algorithm presented in Algorithm 3.28.1. In this version a randomized step is replaced with a deterministic one, based on a solution of a linear programming formulation of an auxiliary problem associated with FAS in bipartite tournaments—namely relaxation of the linear programming

---

Algorithm 3.28.1 Rand-MFASBT($T = (V = (V_1, V_2), A)$)—randomized algorithm
for FAS in bipartite tournament

Step 0: If T is a directed acyclic graph then run the topological sort algorithm on T and let X
be the order of V returned by the algorithm. **return**(X).
Step 1: Randomly select an arc $(i, j) \in A$, where $i \in V_1$, $j \in V_2$.
Step 2: Form the 2 sets V_L and V_R as follows.
$\quad V_L = \{u \mid (u, i) \in A \text{ or } (u, j) \in A\}$
and
$\quad V_R = \{v \mid (i, v) \in A \text{ or } (j, v) \in A\}.$
$\quad$ {It is clear that $T[V_L]$ and $T[V_R]$ are bipartite sub-tournaments of T }
Step 3:
$\quad$ **return**($Rand\text{-}MFASBT(T[V_L]), i, j, Rand\text{-}MFASBT(T[V_R])$)

---

formulation for a 4-cycle Hitting problem. Pseudo-code for the aforementioned
deterministic version of the method can be seen in Algorithm 3.28.2.

Algorithm 3.28.2 Det-MFASBT($T = (V = (V_1, V_2), A)$)—deterministic algorithm
for FAS in bipartite tournament

Step 0: If T is a directed acyclic graph then run the topological sort algorithm on T and let X
be the order of V returned by the sorting algorithm. **return**(X).
Step 1: Select an arc $q = (i, j) \in A$, such that $\dfrac{|R_q|}{\sum_{a \in R_q} x_a^*}$ is minimized.
Step 2: Form the 2 sets V_L and V_R as follows.
$\quad V_L = \{u \mid (u, i) \in A \text{ or } (u, j) \in A\}$
and
$\quad V_R = \{v \mid (i, v) \in A \text{ or } (j, v) \in A\}.$
Step 3:
$\quad$ **return**($Det\text{-}MFASBT(T[V_L]), i, j, Det\text{-}MFASBT(T[V_R])$)

This Det-MFASBT procedure is a deterministic factor-4 approximation algorithm for Feedback Arc Set in bipartite tournaments. That is, the size of the feedback
arc set returned by the algorithm is at most 4 times the size of a minimum feedback
arc set of T. By a simple modification, this algorithm can be generalized for
multipartite tournaments. When an arc $q = (i, j) \in A$ is picked as a pivot,
a partition containing i is treated as one partition, and union of all other parts
as another one, making it "similar" to a bipartite tournament. In this way two
smaller instances of multipartite tournaments are created, and now the algorithm can
separately recurse. The analysis for bipartite tournaments can now be carried over
for multipartite tournaments, therefore the modified Det-MFASBT(T) is a factor-4
approximation algorithm for FAS problem in T.

It has been later on determined that the algorithm of the paper in question does
not give a constant-factor approximation algorithm since a problem in the analysis
of the algorithm has been detected, this has been discovered in [135]. However, an
alternate algorithm was devised for which the result claimed by Gupta in [64] indeed
holds—for a summary of the research found in [135] one can consult Sect. 3.38.

Notwithstanding aforementioned, the algorithm given by Gupta in the paper in question is valuable as an approach for solving FAS, and is therefore included in this historical narrative.

3.29 Aggregating Inconsistent Information: Ranking and Clustering [3]—October 2008

Here, the problems of Rank Aggregation, Feedback Arc Set on tournaments, Correlation-Clustering, and Consensus Clustering are addressed. The authors have found a remarkably simple, and essentially the same, approximation algorithm for all these problems, and various weighted versions of them. Using this algorithm they have obtained improved approximation factors.

The context of the research is the problem of Aggregating Inconsistent Information from many different sources that arises in numerous contexts and disciplines. The goal then is to find a globally consistent solution that minimizes the extent of disagreement with the respective inputs. This Rank Aggregation is NP-Hard [15] even when there are only four input lists to aggregate [45], and is a special case of weighted FAS-Tournament instance.

The algorithm used for FAS-Tournament, called KwikSort, firstly picks a random vertex i to be the "pivot" vertex. Then all vertices connected to i with an in-edge are placed on the left side of i, and all vertices connected to i with an out-edge are placed on the right side of i. Finally, on the two tournaments induced by the vertices on each side a recursion in made. The analysis of this algorithm gives a 3-approximation algorithm for FAS-Tournament—which is an improvement over the best-known previous factor of $O(\log n \log \log n)$. One can refer to Table 3.2 for a summary of approximation factors for different scenarios with combinatorial algorithms. Take note that the authors have also considered Linear Programming (LP) relaxations for FAS-Tournament and Correlation-Clustering. Where after choosing a pivot vertex (instead of in a deterministic way placing vertices on the right or left side), a random decision is made based on LP values. This in turn results in vastly improved approximation factors.

Analysis, results of which can be seen in Table 3.2, has been conducted on various cases of weighted FAS-Tournament, and weighted Correlation-Clustering. Cases analyzed are as follows:

1. **Probability constraints**: $w_{ij} + w_{ji} = 1$ (respectively, $w_{ij}^+ + w_{ij}^- = 1$) $\forall\, i, j \in V$
2. **Triangle inequality**: $w_{ij} \leq w_{ik} + w_{kj}$ (respectively, $w_{ij}^- + w_{jk}^- \leq w_{jk}^-$) $\forall\, i, j, k \in V$
3. **Aggregation**: Edge weights are a convex combination of actual permutations (respectively, clusters). Constraints (1) and (2) are implied in this case.

Term (i, j, k) is used to denote the directed triangle $(i \rightarrow j, j \rightarrow k, k \rightarrow i)$. As for the unweighted instances, given (V, w) of weighted FAS-Tournament, the

Table 3.2 Approximation factors for different scenarios with combinatorial algorithms[a]

Scenario	Ordering	Clustering	Ordering-LP	Clustering-LP
Unweighted instances[b]	3^c	3 (4)	5/2	5/2
Probability constraints *(1)*	5^c	5 (9)	5/2	5/2
Triangle inequality *(2)*	2^c	N/A[d]		
Probability constraints + Triangle inequality *(1,2)*	2^c	2 (3)	2	2
Aggregation *(3)*	11/7 (2)	11/7 (2)	4/3	4/3

[a] Previous best-known factors are shown in brackets
[b] Unweighted majority tournament that corresponds to the input weighted tournament
[c] The best-known factor was the $O(\log n \log \log n)$ algorithm for digraphs [50, 121]
[d] The authors techniques cannot be directly applied to weighted Correlation-Clustering with triangle inequality, but no probability constraints

unweighted majority tournament $G_w = (V, A_w)$ is defined as follows: $(i, j) \in A_w$ if $w_{ij} > w_{ji}$. If $w_{ij} = w_{ji}$, then $(i, j) \in A_w$ or $(j, i) \in A_w$ is decided arbitrarily.

For a minimum FAS on tournaments the authors have devised a randomized algorithm with expected cost at most three times the optimal cost. This algorithm can be seen in Algorithm 3.29.1, and a more concise version, from [97], can be seen in Algorithm 3.29.2.

Algorithm 3.29.1 KwikSort($G = (V, A)$)—algorithm for minimum FAS in tournaments

---

if $V = \emptyset$ **then**
    return empty-list
end if
set $V_L \rightarrow \emptyset$, $V_R \rightarrow \emptyset$
pick random pivot $i \in V$
for all vertices $j \in V \setminus \{i\}$ **do**
    if $(j, i) \in A$ **then**
        add j to V_L {place j on left side}
    else if $(i, j) \in A$ **then**
        add j to V_R {place j on right side}
    end if
end for
let $G_L = (V_L, A_L)$ be tournament induced by V_L
let $G_R = (V_R, A_R)$ be tournament induced by V_R
return order $KwikSort(G_L), i, KwikSort(G_R)$ {concatenation of left recursion, i, and right recursion}

Algorithm 3.29.2 KwikSort(*vertices* S)—algorithm for minimum FAS in tournaments

---

choose a vertex v uniformly at random from S
let L be the set of vertices u such that $w_{uv} \geq w_{vu}$ and $R = S \setminus L$
return the concatenation of $KwikSort(L)$ and $KwikSort(R)$

If one deals with weighted FAS-Tournament instance (V, w) where $w \in (\mathbb{R}^+)^{n(n-1)}$, then construct the unweighted majority tournament $G_w = (V, A_w)$ and return the ordering generated by KwikSort(G_w)—if there exists a constant $\alpha > 0$ such that $w(t) \leq \alpha c^*(t)$ for all $t \in T$, then $E[C^{KS}] \leq \alpha C^{OPT}$, that is KwikSort($G_w$) is an expected α-approximation solution. For more details one should consult section 4, Minimum Feedback Arc Set in Weighted Tournaments, of the paper in question. Running algorithm KwikSort on G_w gives an expected 5, and 2, approximation for the probability constraints case, and the triangle inequality constraints case, respectively.

In order to round LP one can use a pivot scheme. The main idea is that after choosing a certain pivot LP solution variables are used in order to randomly decide where to put all other vertices (instead of deciding greedily). The LP-based algorithm is solving LP only once, and then use the same LP solution in all recursive calls. The LP-based algorithm for weighted FAS-Tournament can be seen in Algorithm 3.29.3, and an LP relaxation can be seen in Eq. 3.6.

Algorithm 3.29.3 LP-KwikSort(V, x)—a recursive algorithm for rounding the LP for weighted FAS-Tournament

---

Input: Given an LP solution, $x = \{x_{ij}\}_{i,j \in V}$.
Output: Returns an ordering on the vertices.

if $V = \emptyset$ **then**
    return empty-list
end if
pick random pivot $i \in V$
set $V_L = \emptyset$, $V_R = \emptyset$
for all $j \in V$, $j \neq i$ **do**
    with probability x_{ji}
    add j to V_L
    else with probability $x_{ij} = 1 - x_{ji}$
    add j to V_R
end for
return order $LP\text{-}KwikSort(V_L, x), i, LP\text{-}KwikSort(V_R, x)$

$$min \sum_{i<j} (x_{ij} w_{ji} + x_{ji} w_{ij}) \; s.t.$$

$$x_{ik} \leq x_{ij} + x_{jk} \; \forall \; distinct \; i, j, k \tag{3.6}$$

$$x_{ij} + x_{ji} = 1 \; \forall i \neq j$$

$$x_{ij} \geq 0 \; \forall i \neq j$$

3.30 Minimum Feedback Arc Sets in Rotator Graphs [90]—January 2009

The authors of this research have presented two algorithms for the problem of FAS, both of which construct FAS for a rotator graph with $n!$ nodes in $O(nn!)$ time. FAS given by the algorithms is minimum in size. Rotator graph is represented by a directed permutation graph which has $n!$ vertices and n sub-graphs, where n represents a graph scale—therefore a graph of scale n is denoted as n-rotator graph [34]. Edges of sub-graphs can be undirected. Pseudo-code for the FAS algorithm on a rotator graph whose subgraph edge is directed can be seen in Algorithm 3.30.1.

Algorithm 3.30.1 FAS-Rotator-I—FAS_n is a FAS in an n-rotator graph

---

for all node $v = s_1 s_2 s_3 \ldots s_n$ in an n-rotator graph **do**
   $flag = true$
   for $i = 2$ **to** n **do**
      if $s_1 < s_i$ **and** $flag$ **then**
         add (v, g_i) to FAS_n
      else
         $flag = false$
      end if
   end for
end for

Rotator graph has $n!$ nodes, with this in mind, Algorithm 3.30.1 is executing a loop $n!$ times and checks all outgoing arcs for each node in this loop. If any of the outgoing arcs satisfy a condition, the arc is added to the feedback arc set. The size of the FAS for an n-rotator graph given by Algorithm 3.30.1 is $\sum_{2 \le i \le n} n!/i$, and is minimum in size, as stated by theorem 3 of the paper in question.

If one has a rotator graph where subgraph edge is undirected, then Algorithm 3.30.1 needs to be modified. Such a modified algorithm for finding FAS in rotator graph is presented in Algorithm 3.30.2. This algorithm is generating FAS for an n-rotator graph of a size $(n!/6 + \sum_{3 \le i \le n} n!/i)$, and this feedback arc set is also minimum in size, just as it was for the previous algorithm. Both algorithms have the same running time, which is stated above, and they are both outputting optimal solution.

There is also a paper published later on in an International Journal of Foundations of Computer Science, by the same set of authors, in which one can additionally find a concise formula for finding a minimum FAS in an incomplete rotator graph. This follow-up research is found in [90].

Algorithm 3.30.2 FAS-Rotator-II—FAS_n is a FAS in an n-rotator graph

for all node $v = s_1 s_2 s_3 \ldots s_n$ in an n-rotator graph **do**
  if $s_1 < s_2$ **then**
    $flag = true$
    Phase 1:
    for $i = 3$ **to** n **do**
      if $s_1 < s_i$ **and** $flag$ **then**
        add (v, g_i) to FAS'_n
      else
        $flag = false$
      end if
    end for
    Phase 2:
    if $s_1 < s_2 < s_3$ **then**
      add (v, g_2) to FAS'_n
    end if
  end if
end for

3.31 Ranking Tournaments: Local Search and a New Algorithm [31]—May 2009

Ranking is a fundamental activity for organizing and understanding data. This ranking can be looked at as a form of advice. If this advice is consistent and complete, then there is a total ordering on the data and the ranking problem is a sorting problem. If on the other hand we have an advice that is consistent, but is incomplete, then ones speaks about topological sorting. And if one has an advice that is inconsistent, then we are speaking about the problem of FAS where one is trying to satisfy the advice given as much as possible. When advice is given about every pair of items we are dealing with a tournament.

The main thrust of the paper in question is reexamining a certain number of existing algorithms and developing some new techniques for solving FAS. These algorithms are tested on both synthetic, and non-synthetic datasets. It has been shown that in practical situation local search algorithms present a powerful tool, even though they are without approximation guarantee.

A new algorithm that the authors have designed is based on reversing arcs, in an organized manner, whose nodes have large in-degree differences, to destroy directed triangles which eventually leads to a total ordering. Combining this approach with a powerful local search yields an algorithm that is on a par, or stronger, than existing techniques on a variety of data sets.

In section 2 of the paper in question, Algorithm Details, the authors are giving an overview of the algorithms they have investigated, and their own improvements. Considering that the aim of this book is algorithmic in nature we will repeat these algorithms here. This will give further insight not only into the paper in question, but to the body of knowledge as well. Kendall score-based algorithms are as follows.

Algorithm *Iterated Kendall* [82] Rank the nodes by their Kendall scores (the Kendall score of a vertex is its in-degree), with lowest Kendall score on the left. If there are nodes with equal score, break ties by recursing on the subgraph defined by these nodes. If there is a subgraph whose elements all have equal in-degree, rank them arbitrarily.

Algorithm *Eades* [47] Select the node that has the smallest Kendall score and place it at the left, breaking ties arbitrarily. Recurse on the remainder of the nodes, having recomputed the in-degrees on the remaining subgraph.

In order to improve Eades algorithm from [47] the authors have allowed the selection of a vertex to either end of the ranking. If there is a vertex v of extremely low out-degree, but no vertex of low in-degree, it makes sense to place v on the right-hand side of the ordering. Let $In(v)$ stand for the in-degree of node v, with $Out(v)$ for its out-degree, then the improved algorithm is as follows.

Algorithm *Eades Improved* Select the node u that maximizes $|In(u) - Out(u)|$ and place it at the left end, if $In(u) < Out(u)$, otherwise, the right end. Recurse on the subgraph induced by removing u. Again, break ties arbitrarily.

Considering that sorting algorithms provide schemes for deciding which of the advice to believe, they have been a source of inspiration for a number of algorithms for FAS, in spite of the fact that FAS lacks transitivity on which sorting algorithms are based on. A general sorting strategy for FAS can be defined based on some sorting algorithm S: run S over the vertices of G, using as the comparison function, $u < v$ iff $u \to v$. This strategy yields the following Quicksort algorithm.

Algorithm *Quicksort* Choose a pivot $p \in V$, uniformly at random. Let $L \subseteq V$ be all vertices v such that $v \to p$, and let $R = V \setminus (L \cup \{v\})$. Let π_L be the ordering of L obtained by quicksort, and π_R be the analogous ordering of R. Output (π_L, v, π_R), the ordering resulting from placing vertices in L on the left (ordered by π_L), etc.

The authors have also studied the merge sort [23] approach, a well known and popular sorting algorithm often compared to Quicksort [73]. No previous mention of this approach used in such a context have been found in the literature.

A liner ordering problem have seen application of Insertion sort [85] technique that is more involved than the usual approach. This Insertion sort inspired algorithm, named Sort, is defined as follows.

Algorithm *Sort* [25] Make a single pass through the nodes from the left to the right. As each node is considered, it is moved to the position to the left of its current position that minimizes the number of back-arcs (if that number of back-arcs is fewer than its current position).

The algorithm Sort is repeated multiple times, until there can no more be any improvement in the number of back-arcs. If the order of the nodes is reversed, this reversal will not increase the number of back-arcs. The version of the algorithm that utilizes reversing the order of nodes has been demonstrated faster on general graphs.

Another approach useful not only for FAS but also for many other problems is Local Optimization. The core of this procedure is to start with some solution, and then iteratively improve this solution until no further improvements are possible. A general local search scheme considered in the paper in question is as follows.

Algorithm *Local Search* Given some solution π, consider all potential local improvements that could be made to π. Choose as a new order π', the local improvement which minimizes the cost, as long as that cost is strictly less than that of π. If no such improvement exists, return π—otherwise repeat.

It is of course crucial to know which permutation represents a local improvement of some ranking π. The authors have considered two heuristics for the task:

1. **swaps**—which swaps the position of two nodes in the order
2. **moves**—which moves one node to any position in the order, leaving the relative order of the other nodes unchanged.

Neither of the algorithms can provide an approximation guarantee. Preliminary testing showed that move heuristics performed well, while swapping did not, therefore swapping was removed from consideration. While without an approximation guarantee in these instances, it is possible to circumvent this by starting a local optimization with a solution outputted from an approximation algorithm, therefore inheriting such bound.

Algorithm from Chanas et al. [25] has the effect of checking, for each vertex of the graph, from left to right, if a move to the left is possible. The operation mentioned, sort o reverse, is doing the same thing, only with moves to the right. With this in mind the authors have developed an alternative algorithm.

Algorithm *Chanas Both* Run Chanas algorithm, but change the sort procedure, so it is allowed to move a node either left or right, to the position that results in the fewest back-arcs.

Of course, this alternation will as a result have a situation where some nodes may be moved more than once in a single sort pass.

Another way of achieving a solution for the problem at hand are a triangle-destroying algorithms. A tournament has a cycle iff it has a directed triangle (Δ). This is due to the complete nature of a tournament graph—every (non-Δ) cycle must have a chord inside it that forms a strictly smaller cycle. Authors have therefore decided to consider triangle-destroying approach where a directed triangle is destroyed by arc reversal. Arc deletion was not an option considering that in such a way tournament would be destroyed. Scheme of the authors is as follows.

Algorithm *Triangle Reversal* While the tournament is not acyclic, choose an arc and reverse its orientation. Once the tournament is acyclic, use the topological sort of the vertices as the solution to the original problem.

There are various heuristics that could be employed in choosing an arc to be reversed. This choice can of course affect both performance and running time of the algorithm. One heuristic way of choosing an arc could be by triangle count. The triangle count of a tournament is the number of directed triangles (Δ) in that tournament. Therefore an arc can be in a number of these Δ, and this represents its triangle count. The first algorithm for aforementioned is therefore as follows.

Algorithm *Triangle Count* Run triangle reversal, choosing on each iteration the arc with highest triangle count.

This heuristic has an issue tough. Reversing an arc can actually create a new Δ which did not exist before, the problem being a larger cycle surrounding the reversed arc. This means that in order to resolve this, one of the outer arcs need to be reversed, one of the arcs that has larger triangle count. It is, however, also possible that reversing an arc with larger triangle count does not resolve the issue, triangle count of the tournament stays the same, and that arc still has the highest triangle count—leading to an infinite loop. The following algorithm is then defined as follows.

Algorithm *Triangle Delta* Run triangle reversal, choosing the arc that causes the greatest net reduction to the tournament's triangle count.

If T is a tournament, and if T has a cycle, then there exists an arc $e \in T$ such that reversing e will reduce the triangle count of T—this has been proved by the authors and therefore mitigated a potential issue with the triangle delta algorithm.

In experiments where triangle count algorithm finished execution it has tended to outperform the triangle delta algorithm, therefore the authors have decided to merge these two approaches. Algorithm is then as follows.

Algorithm *Triangle Both* Run triangle reversal, choosing the arc with the highest triangle count, provided that it reduces the tournament's total number of Δs.

The best algorithm that the authors had at their disposal for calculating the triangle count for every arc of the digraph required $O(n^3)$ operations. If one has a weighted tournament, then the weight of Δ is the sum of the weights of its arcs. Consequently, the triangle count of an arc is the sum of the weights of the Δs an arc is involved in.

The authors have also designed a new algorithm, a degree difference algorithm. This algorithm is also a triangle reversal algorithm, but it uses a much simpler computation than is a computation for a full triangle count. The algorithm designed is as follows.

Algorithm *Degree Difference* Run triangle reversal, choosing the arc $u \rightarrow v$ for which the difference between u's in-degree and v's in-degree is greatest.

It may take $\Theta(n)$ time to find such an arc for each iteration, but this algorithm always make progress toward total order. In order to speed up this approach the authors have employed a sampling technique. Sampling $\log n$ vertices favoring high in-degree, and the same for favoring low in-degree. Then the algorithm checks each of the $\log^2 n$ arcs between sampled vertices, choosing the back-arc of highest degree difference—if back-arcs only of non-positive degree difference were found, then re-sampling is employed. This algorithm, named Degree Difference Sampled 1, takes $O(n^2 \log^2 n)$ time on average.

Another variation, Degree Difference Sampled 2, maintains two lists. One list of potential head nodes, and one of potential tail nodes. The idea being pushing the quantity $\sum_v In(v)^2$ toward its maximum (which is reached when all in-degrees are different). A node v is a potential head if its in-degree is not unique, or there is no node of in-degree $In(v) - 1$. A node u is a potential tail if its in-degree is not unique, or there is no node of in-degree $In(u) + 1$. In this algorithm $\log n$ nodes are sampled from each list uniformly, and then from these pairs an arc to reverse is selected with the largest in-degree difference. For reasons why certain algorithmic approaches cannot guarantee reasonable factor approximations one should consult section 3 of the paper in question, Approximation Counter-Examples.

The research also extends to empiricism, in order to validate practical performance, which was conducted on a 4-core Intel Xeon 3.2 GHz machine, with 8 GB of main memory. All algorithms were compiled by GCC version 3.4.6 with the -O3 optimization flag set. In order to ascertain significance of initial solution quality to the effectiveness of local search techniques, algorithms were firstly tested in isolation, starting from a random ordering, after which results were passed to both the Chanas and moves algorithms.

Presenting as input to Chanas an output of Chanas had surprising consequences. Two calls to the algorithm can lower the cost of a ranking, while one will not, and in this way improve upon a local optimum. This can happen because repeated calls to Chanas can sometimes move the algorithm out of a local plateau.

Datasets upon which experiments were conducted were diverse, and ranged from artificial, as the authors called it, synthetic dataset, to real-word instances of rankings of a large set of documents. The authors have also tested on a dataset where they have generated tournaments of movie rankings that represented the "consensus view" of selected groups.

After conducted testing results were as follows. Chanas local search algorithm performed well, and was rarely beaten. Quicksort and Merge sort procedure were performing poorly. Bubblesort was performing poorly, but if (to its execution) Chanas procedure is applied, then it performed well. Both Eades and Eades improved algorithms performed strongly, although the latter had lower running time. The Triangle Both algorithm performs very slowly, but this changes when local search is combined with it. The Degree Difference algorithm was similar to the Triangle one, although sampling methods for Degree Difference were better compromises.

When the problem size increased Triangle Both and Degree Difference algorithm had very high empirical running times, which was not surprising having in mind their cubic order of growth. The fastest were iterated Kendall, Quicksort, Chanas both, and Eades improved. The conclusion being that improvements the authors made were delivering speed increase. In the middle of the pack were Chanas, Eades, and Degree Difference Sampled 1 (which runs in similar time to Chanas, with an advantage on larger datasets). Degree Difference Sampled 2 and moves algorithms were faster than Triangle Both and Degree Difference, but slower than the rest of the algorithms. For a more detailed look one should consult section 5 of the paper in question, Discussion of Results.

3.32 Deterministic Pivoting Algorithms for Constrained Ranking and Clustering Problems [138]—August 2009

In this paper we find ranking and clustering related to the aggregation of inconsistent information: Rank Aggregation, (weighted) Feedback Arc Set in tournaments, Consensus and Correlation-Clustering, and Hierarchical Clustering. This work is an extension of the work in [3] where randomized constant-factor approximation algorithms were devised—these were recursively generating a solution by randomly choosing a pivot, and dividing the remaining vertices into two groups based on the pivot vertex.

In this extended work the authors have given deterministic approximation algorithms for the aforementioned problems, which were lacking in previous research. The authors have also considered the problem of finding minimum-cost ranking and clustering with constraints (e.g., partial order input for the ranking problem)—these constrained problems can be solved by the algorithm the authors have designed. As for the Rank Aggregation or Consensus Clustering problem, if input obeys the constraints, then the output of devised algorithms will obey the constrains without increasing the objective value of the solution.

In the continuation we will of course have a focus on the problem of FAS. The algorithms are simple, derandomized versions of combinatorial algorithms in [3], where algorithms analysis implies the performance guarantees for the original algorithms as well. Devised deterministic algorithms are adapted in order to tackle constrained problems as well. The authors have also extended ideas from their own deterministic algorithms to the randomized rounding algorithms from [2, 3], and have showed how to derandomize these also—the analysis here again implies approximation guarantees of the original randomized algorithms.

The authors have proved the existence of a deterministic combinatorial 2-approximation algorithm for weighted feedback arc set in tournaments with triangle inequality, which runs in $O(n^3)$ time. As pseudo-code is not given, and without repeating large parts of the paper, for more details one should look into theorem 2.1, lemma 2.1, and theorem 2.2 of the paper in question. The basis of the procedure is a framework which can be used to obtain a solution to a given input of a weighted Feedback Arc Set problem, pseudo-code can be seen in Algorithm 3.32.1, which has its origin in [3].

Algorithm 3.32.1 FAS-Pivot($G = (V, A)$)

---

pick a pivot $k \in V$
$V_L = \{i \in V : (i, k) \in A\}, V_R = \{i \in V : (k, i) \in A\}$
return $FAS\text{-}Pivot(G(V_L)), k, FAS\text{-}Pivot(G(V_R))$

In the previously published algorithm [3] $G = (V, A)$ is a majority tournament, and a pivot is being chosen randomly. Here, however, G is either the majority tournament, or a tournament obtained by rounding the solution to a linear programming relaxation. While choice of a pivot will be constrained with a lower bound on the weight of an optimal solution, this will ensure that a "good" pivot is chosen.

If one only knows that the weights satisfy probability constraints, then one needs to turn to linear programming relaxation which will provide a more robust lower bound. With such a linear program one can achieve a deterministic 3-approximation algorithm for weighted minimum Feedback Arc Set with probability constraints—for details one should consult theorem 2.4 of the paper in question. It is also possible to use LP-based method to obtain a 2-approximation algorithm for instances when the weights satisfy the triangle inequality—by a simple modification detailed in remark 2.2 of the paper in question.

$$A_c = \frac{\mathbb{E}\left[W_k(V)|V_L \cup \{i\}, V_R\right]}{\mathbb{E}\left[C_k(V)|V_L \cup \{i\}, V_R\right]}$$

$$B_c = \frac{\mathbb{E}\left[W_k(V)|V_L, V_R \cup \{i\}\right]}{\mathbb{E}\left[C_k(V)|V_L, V_R \cup \{i\}\right]} \tag{3.7}$$

Results for weighted problems with probability constraints also hold for constrained versions of these problems. The authors have therefore showed that there exists a deterministic 3-approximation algorithm for constrained weighted minimum Feedback Arc Set with probability constraints, and that there exists a 2-approximation algorithms for ranking with triangle inequality based on linear programming relaxation.

$$C = \frac{\mathbb{E}\left[\sum_{(i,j)\in T_k(V)} w_{(i,j)}\right]}{\mathbb{E}\left[\sum_{(i,j)\in T_k(V)} c_{ij}\right]} \tag{3.8}$$

In order to derandomize the randomized rounding algorithms in [2, 3] authors have extended ideas which they used thus far. This enabled deterministic $\frac{5}{2}$-approximation algorithm for Ranking with probability constraints, a $\frac{3}{2}$-approximation algorithm for Ranking with triangle inequality, and $\frac{4}{3}$-approximation algorithm for full Rank Aggregation. The pseudo-code for the randomized version of the algorithm for weighted FAS can be seen in Algorithm 3.32.2, while derandomized version presents itself in Algorithm 3.32.3.

In Algorithm 3.32.3 terms A_c and B_c are defined as in Eq. 3.7, and term C is defined as in Eq. 3.8—where $\mathbb{E}$ denotes expected cost.

Algorithm 3.32.2 FASLP-Pivot(V, p)

---

pick a pivot $k \in V$
set $V_L = \emptyset$, $V_R = \emptyset$
for all $i \in V, i \neq k$ **do**
    with probability $p_{(i,k)}$: add i to V_L
    else with probability $p_{(k,i)}$: add i to V_R
end for
return $FASLP\text{-}Pivot(V_L, p), k, FASLP\text{-}Pivot(V_R, p)$

Algorithm 3.32.3 DerandFASLP-Pivot(V, p)

---

pick $k \in V$ minimizing C
set $V_L = \emptyset$, $V_R = \emptyset$, $V' = V \setminus \{k\}$
while there exists $i \in V'$ **do**
  if $A_c \leq B_c$ **then**
    add i to V_L
  else
    add i to V_R
  end if
  $V' = V' \setminus \{i\}$
end while
return $DerandFASLP\text{-}Pivot(V_L, p), k, DerandFASLP\text{-}Pivot(V_R, p)$

3.33 Faster FAST (Feedback Arc Set in Tournaments) [51]—November 2009

This research has resulted in the algorithm that finds a feedback arc set of size k in a tournament (k-FAST), and raises a question if tournament has such a feedback arc set, with running time being $n^{O(1)} + 2^{O(\sqrt{k})}$, is a continuation of the work in [4], where n stands for the number of vertices in the tournament and k represents tournament size. The running time of this algorithm was asymptotically faster than running time of previously known algorithms for this problem, and remains polynomial for $k = O((\log n)^2)$.

The algorithm of the paper in question is based on Dynamic Programming, but it does not uses random coloring on which algorithm in [4] was based. The general steps for the algorithm can be seen in the continuation.

1. **Compute** the set B of **bad vertices**.
2. For each location i in the linear order, **compute a candidate set** $C(i)$ that contains those vertices whose in-degree is between $i - d$ and $i + d$, plus the vertices of B. In addition, **compute a prefix set** $P(i)$ that contains those vertices not in B with in-degree less than $i - d$.
3. Using these candidate sets and prefix sets, **compute a minimum feedback arc set** using Dynamic Programming.

Aforementioned set of bad vertices B consists of vertices where at least t arcs incident with it are major suspects, while an arc is a suspect if it belongs to some triangle and a major suspect if it belongs to at least t triangles.

With Dynamic Programming one can find which linear order, among those that respect the candidate sets, has the smallest FAS, by scanning locations from 0 to $n - 1$. When one reaches location i he needs to know the following:

1. Which **vertices** of $C(i)$ **have been placed up to location** i. There are at most $2^{|C(i)|}$ possibilities for such subsets.
2. How many **backward arcs** we have **placed so far**. For each choice of subset $C'(i)$ as in item 1, we need to remember just the smallest number of backward arcs, attained in a linear arrangement, up to i placed $C'(i)$ and did not place $C(i) - C'(i)$.

If $C(i) - C'(i)$ is empty, then corresponding branch of the dynamic programming ends. $C'(i + 1)$ can be computed as $C(i + 1) \cap (C'(i) \cup \{v\})$, and the number of backward arcs can be easily updated by summing to the previous total the arcs from v to $C'(i)$ and from v to $P(i)$.

3.34 Fixed-Parameter Tractability Results for Feedback Set Problems in Tournaments [42]—March 2010

The authors that have published this paper have reported improvement in fixed-parameter tractability for Feedback Set problems. More concretely, concerning FAS, a depth-bounded search tree for FAS in bipartite tournaments (FASBT), based on a new forbidden subgraph characterization with running time being $3.373^k \cdot n^{O(1)}$.

Following the same approach as in [113] the authors have designed a fixed-parameter algorithm for Feedback Arc Set in bipartite tournaments. The research of the paper in question is contrasted to [113] by the fact that here two sub-tournaments are needed for instance characterization, alongside with a more involved branching strategy. A bipartite tournament is acyclic iff it contains no cycle of length four—it should be noted that if a bipartite tournament B contains neither B_1 nor B_2 as an induced subgraph, then all cycles in B have length four and are pairwise vertex-disjoint (see Figs. 3.2 and 3.3).

The authors algorithm for Feedback Arc Set in tournaments, of the paper in question, therefore has the same two phases as the algorithm in [113]. More precisely, this algorithm can solve FASBT of n vertices with k arc deletions in $O(3.373^k \cdot n^3)$ time, where phases for the algorithm are as follows:

1. Search tree procedure **destroying all cycles** contained in the induced sub-graphs B_1 and B_2
2. Polynomial-time procedure **getting rid of** the remaining **vertex-disjoint cycles**.

Fig. 3.2 Forbidden subgraph B_1 for bipartite tournaments (all cycles of length four are disjoint), color of the vertices describes bi-partition

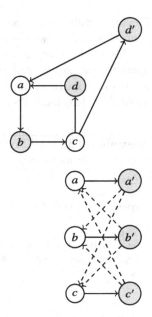

Fig. 3.3 Forbidden subgraph B_2 for bipartite tournaments (all cycles of length four are disjoint), color of the vertices describes bi-partition

3.35 Fast Local Search Algorithm for Weighted Feedback Arc Set in Tournaments [56]—July 2010

In this paper one can find a parameterized sub-exponential time local search algorithm which finds an improved solution, if there is any, in the k-exchange neighborhood of the given solution to an instance of weighted Feedback Arc Set in tournaments. Given an arc-weighted tournament T on n vertices and a feedback arc set F, the algorithm that the authors have devised decides in $O(2^{o(k)}n \log n)$ time if there exists a feedback arc set of smaller weight, and that differs from F in at most k arcs. It seems that this is the first algorithm searching the k-exchange neighborhood of an NP-Complete problem that runs in parameterized sub-exponential time.

As an interesting side note, using this local search algorithm for weighted FAST the authors have obtained sub-exponential time algorithms for Kemeny ranking and one sided Crossing Minimization, a problem in graph drawing, which "consists of placing the vertices of one part of a bipartite graph on prescribed positions on a straight line and finding the positions of the vertices of the second part on a parallel line and drawing the edges as straight lines such that the number of pairwise edge crossings is minimized." [101] Fast search of the k-exchange neighborhood is also possible for dense structures, as the authors have showed.

By using chromatic coding from [4] the authors of the paper in question presented k-LSFAST parameterized algorithm which can by exchanging a number of arcs from the current solution solve FAST randomly in time $2^{O(\sqrt{k}\log k)}n$, where deterministic version is also possible and runs in time $2^{O(\sqrt{k}\log k)}n \log n$. Even though this algorithm is similar to the one in [4] there is a crucial difference.

Previously published algorithm starts by preprocessing the instance and obtains an equivalent instance with at most $O(k^2)$ vertices in polynomial time.

This preprocessing allows them to assume that the instance where they actually apply the sub-exponential time algorithm is of size $O(k^2)$ only, which is integral to their time analysis. Here on the other hand, the authors have resorted to a more sophisticated Dynamic Programming strategy.

Algorithm 3.35.1 k-LSFAST—outline of the algorithm for k-exchange neighborhood weighted FAS in tournament

---

Step 1

let $t = 4\sqrt{k}$
color the vertices of T uniformly at random with colors from $\{1, \ldots, t\}$

Step 2
let A_c be the set of arcs whose endpoints have different colors
find a minimum weighted feedback arc set $F' \in \mathcal{N}_k(F)$ such that S_{in} and S_{out} are contained in A_c

---

For an outline of the algorithm one can look at Algorithm 3.35.1—$\mathcal{N}_k(F)$ denotes set of neighbors of F with respect to k-exchange neighborhoods, with S_{in} and S_{out} representing arc subsets where $F' = F \setminus S_{out} \cup S_{in}$. The probability that G is properly colored is at least $(2e)^{-\sqrt{q/8}}$.

Algorithm 3.35.1 named k-LSFAST can be derandomized, by using universal coloring families from [4]. A family $\mathcal{F}$ of functions is called a universal coloring family if for any graph G on the set of vertices with at most k edges, there exists an $f \in \mathcal{F}$ which is a proper vertex coloring of G. Therefore an explicit construction of a coloring family can replace the randomized coloring step of the algorithm in question, which yields a deterministic sub-exponential time algorithm for k-LSFAST. For various applications of the algorithm that the authors are directly tackling one should consult the paper itself, under a section 3, in a subsection titled, Other Applications.

3.36   Faster Algorithms for Feedback Arc Set Tournament, Kemeny Rank Aggregation and Betweenness Tournament [80]—December 2010

Authors of this paper have presented deterministic fixed-parameter algorithm for weighted FAS on tournament. The algorithm has a time complexity of $O^*(2^{O(\sqrt{OPT})})$, which represents an improvement over the algorithm in [4]—$O^*(\cdot)$ hides polynomial factors and parameter OPT represents the cost of an optimal ranking. Pseudo-code for the algorithm can be seen in Algorithm 3.36.1, with $r(v)$ being computed uncertainties in the positions of all the vertices v. The lower bound, as per exponential time hypothesis in [76], for a run-time is $O^*(2^{o(\sqrt{OPT})})$. The authors have reported that problem instances with OPT being as large as $n(\log n)^2$ could be solved in polynomial time.

Algorithm 3.36.1 Exact FAST—for both Dynamic Programming and Divide-and-Conquer techniques

---

Input: Vertex set V, arc weights $\{w_{uv}\}_{u,v \in V}$.
Output: Found optimal ranking.

sort by weighted in-degree [33], yielding ranking π^1 of V
set $r(v) = 4\sqrt{2C(\pi^1)} + 2b(\pi^1, v, \pi^1(v))$ for all $v \in V$
use Dynamic Programming or Divide-and-Conquer to find the optimal ranking π^2 with $|\pi^2(v) - \pi^1(v)| \leq r(v)$ for all v

---

By solving a Dynamic Program, or a Divide-and-Conquer variant, for the optimal ranking the algorithm outputs a solution near a particular constant-factor approximate ranking—with the variant needing $n^{O(\sqrt{OPT})}$ time. Before the algorithm is run a preliminary step is conducted. A small kernel is computed, a smaller instance that is, with the same optimal cost as the input instance. This preliminary step allows separation of dependence on n and OPT during run-time. This $O(OPT^2)$-vertex kernel for weighted FAST has been computed, with [41] being a foundation, in polynomial time.

In order to compute the kernel two reduction rules are applied as often as possible. The first reduction rule is eliminating a vertex that is part of no cycles of three arcs in the majority tournament. The second reduction rule being concerned with an arc (u, v) of the majority tournament which is in more than $2U$ cycles of three arcs in the majority graph, where $OPT \leq U \leq 5OPT$ is the cost of a 5-approximate ranking [33].

Any feedback arc set that is not paying for such an arc must pay for more than $2OPT$ other arcs of the majority tournament (with each costing at least $1/2$) and hence cannot be optimal. Therefore it is recorded that one must pay w_{uv}, consequently setting w_{uv} to zero and w_{vu} to one. If one is interested in a Betweenness tournament, then consult section 3, Betweenness tournament, for details.

3.37 Sorting Heuristics for the Feedback Arc Set Problem [21]—February 2011

Research presented in the paper in question deals with the problem of FAS in the context of extending classical sorting algorithms to heuristics for the problem of FAS, with directed arcs serving as binary comparators. The algorithms have been analyzed, but new hybrid algorithms have also been developed so as to gain further improvements. The performance between various algorithms is similar and varies by about 0.1%, with primary difference being the use of Insertion sort and Sifting. Running time and convergence to a local minimum also represents a point of difference. Generally speaking, conducted experiments have showed that Sifting is a better performer than Insertion sort.

Algorithm in [3], termed KwikSort, has been extended by the authors to arbitrary directed graphs, while polynomial time approximation scheme from [83], which allows the computation of a feedback arc set with a size of at most $1 + \epsilon$ times the optimum for any $\epsilon > 0$ (by applying KwikSort once and improving the arrangement by repeatedly removing vertices from the ranking and reinserting them at another position), equals authors Sifting algorithm.

If input for the algorithm is not simple directed graph, then some preprocessing steps need to be taken. Each self-loop has no impact on other cycles in the graph. Therefore self-loops are firstly removed from the graph, but are later on added to FAS. Two-cycles, (u, v) and (v, u), can be processed by first erasing both arcs and then adding into the FAS that one which conforms to the linear arrangement. Parallel arcs are transformed into a bundle of paths, of length two, by splitting each of them by an auxiliary vertex. A digraph can be decomposed into its strongly connected components, which can in turn be topologically sorted—FAS must respect the order of arcs (u, v) if u and v belong to different strongly connected components, and u must be listed before v in the topological order (by this being true the linear arrangement is SCC-conform). It is time consuming to repeatedly compute strongly connected components and their topological order, while the loss in general is small.

In a post-processing phase, one is to clean up the obtained FAS, or a linear arrangement π. A set of arcs is minimal FAS if reinsertion of any arc $f \in F$ induces a cycle. This can be done in $O(|F|) \times O(n + m)$ time, while for the dense graphs with large feedback arc sets it comes to $O(n^4)$—which is costly (n: vertices; m: arcs). If vertices u and v are adjacent in a linear arrangement π, then u must appear before v if there is an arc (u, v), otherwise they are swapped—this can be done in $O(n^2)$ time.

An algorithm is said to be monotone if the output is never worse than the input thereof. In such a situation the size of FAS is never increased, it is therefore promising to run the algorithm multiple times. Hence the convergence number for a monotone algorithm A is the number of runs of A that strictly improve the result. Such a general algorithm can be seen in Algorithm 3.37.1, for more details one should look at lemma 1 of the paper in question.

Algorithm 3.37.1 A^*

---

    function $Iterate$(algorithm A, cost c, arrangement π)
    repeat
        $\pi' \leftarrow \pi$
        $\pi \leftarrow A(\pi)$
    until $c(\pi) \geq c(\pi')$
    end function

Table 3.3 Analysis of sorting heuristics

Heuristic	Running time	Gap to optimum [31]
ELS[a]	$O(n+m)^{\text{e}}$	$O(n)$
ELS-abs[a]	$O(n+m)^{\text{e}}$	$O(n)$
KS3[b]	$O(n \log n)^{\text{f}}$	
Sort[c,g]	$O(n^2)$	$O(n)$
Sifting[d,g]	$O(n^2)$	$O(n)$

[a] Based on [47], alternate version used in [31]
[b] Based on [3]
[c] Based on [25]
[d] For example, used in [99]
[e] With an upper bound of $m/2 - n/6$ for the size of FAS
[f] An average running time
[g] Monotone

The result of algorithm A* is a first local optimum for an algorithm A. If A is not monotone, then the result of the next to last run should be taken. Enforcement of a strict improvement generally leads to a faster termination of the algorithm and a lower convergence number. One could also repeat A until it worsens the result for the first time. Commonly used 1-OPT property could be used, where the removal of an item and its reinsertion does not lead to an improvement. The exchange of vertices, that is Sifting [11], for the feedback arc set could be used, the exchange being called 2-OPT.

The algorithms that have through modifications been considered by the authors are in particular based on Selection sort, Insertion sort and Quicksort. Since pseudo-codes have not been given, we will not be repeating large parts of the text from the paper in question, but will refer the reader to the paper itself, section 2.4, Basic Algorithms. These algorithms are ELS and ELS-abs, Quicksort $KS3$, Insertion sort, and Sifting (which can be regarded as a two-sided Insertion sort). For analysis details see Table 3.3.

Hybrid algorithms have been devised in order to overcome some weak points of existing algorithms and so as to merge good properties of these existing algorithms. These hybrid algorithms, however, have higher running time, so a trade-off exists. In the worst case, repetitions of algorithm A to A* (repeated application of an algorithm up to its convergence number) increase the run-time by a factor of $O(m)$, which is a consequence of the bound on the size of the FAS. The following hybrid algorithms were tested:

1. It-**Sort** = Sort*
2. It-**Sift** = Sift*
3. It-**Move** = Move*—called Chanas-Both in [31]
4. **CK-Sort** = Sort* ∘ (Reversal ∘ Sort*)*
5. **CK-Sift** = Sift* ∘ (Reversal ∘ Sift*)*
6. It-**2-Sift** = (Sift* ∘ (Sift$_r$ ∘ Sift*))*
7. **X-Sift** = Sift* ∘ ((Reversal ∘ Sort) ∘ Sift*)*

For example, X-Sift first applies sifting until the convergence number is reached, afterwards it reverses the arrangement and runs sort once, followed by sifting (until sifting reaches its convergence number). As stated in the expression, the latter is repeated up to the convergence number of the combined methods. All hybrids are monotone, but they are not 1-OPT.

Algorithms were tested on several suites (the paper in question reports testing on two data sets), first of which consists of random graphs according to the Erdős and Rényi model $G(n, p)$, where the size of n ranges from 100 to 1000 in steps of 100, and an edge between any two vertices is chosen with probability $p = 0.5$, graphs had about 249750 directed edges. Results were computed over 1000 repetitions, tests were conducted for performance (including on counter-example graphs where Insertion sort and Sifting are trapped in a local optimum, provided that starting linear arrangement is unfavorable), running time, and convergence.

Experiments were executed on AMD Phenom II X6 1090T machine with 8 GB of memory. The algorithms themselves were developed and executed in Java SE Run-time Environment, version 1.6 using the Java HotSpot Server VM. The results show quite similar results when performance is looked upon, size of the computed feedback arc sets differ by about 0.1%. This may be an indication of closeness to optimum, but whether that is the case was unknown, with Sifting outperforming Sorting. With convergence in mind, hybrid algorithms present a different behavior, which has a direct impact on the run-time. Sifting was faster than Insertion sort, and pure Sorting and Sifting heuristics were faster than the CK algorithms. For additional details one should consult the paper in question, figures 3, 4 and 5, and tables 1, 2 and 3.

3.38 Linear Programming Based Approximation Algorithms for Feedback Set Problems in Bipartite Tournaments [135]—May 2011

The authors of this paper have considered both Feedback Arc Set and Feedback Vertex Set, on bipartite tournaments. For the problem of FAS, an alternate algorithm to the one presented in [64] has been provided, it is a 4-approximation algorithm based on an algorithm for FAS on (non-bipartite) tournaments presented in [138].

The algorithm presented in [64] is similar to the one proposed in [3], for the Feedback Arc Set problem on tournaments, where it recursively constructs an ordering of vertices—with feedback arc set consisting of arcs going from right to left in the final ordering. The key to the analysis in [64] is the claim that arc $(u, v) \in A$ becomes backward arc iff $\exists (i, j) \in A$ such that (i, j, u, v) forms directed 4-cycle in G and (i, j) was chosen as the pivot when all 4 were part of the same recursive call. Yet, a situation could arise where an arc (u, v) may also become backward if $(i, u) \in A$ and $(v, j) \in A$, and (i, j) was chosen as the pivot when

i, j, u, v were in the same recursive call. In this case there is no directed 4-cycle $\{i, (i, j), j, (j, u), u, (u, v), v, (v, i)\}$ since one has $(i, u), (u, v), (v, j), (i, j) \in A$.

On the other hand, authors of the paper in question, have directly applied derandomization from [138] and have made a more direct extension of the algorithm from [3]. This has allowed them to obtain a 4-approximation algorithm for FAS on bipartite tournaments. Since the authors have not given pseudo-code, and without repeating large parts of the paper, we would direct the reader to the paper itself, section 3, The Feedback Arc Set Problem on Bipartite Tournaments, for more details.

3.39 An Efficient Genetic Algorithm for the Feedback Set Problems [102]—February 2014

In this subsection we will deal with the problem of FAS and a genetic algorithm (GA) proposed for the solution thereof. It seems that drawing of directed graph in a pleasing way was a motivation for the authors work, and especially drawing those that firstly need to be made acyclic—other applications of FAS have also been mentioned.

The authors of the paper in question have also compared their own algorithm with two simple algorithms proposed by Garey and Johnson [59]. Since these two algorithms are of interest in a book such as this one, we will therefore very briefly cover them here.

The first one is a simple heuristic which is assigning unique labels to nodes, $\delta(u) \in \{1, 2, \ldots, |V|\}$, for each node v in a digraph. This is called unique node labeling and an arc $u \to v$ with $\delta(u) < \delta(v)$ is termed an up-arc. The set of these up-arcs therefore forms a feedback arc set. The problem of FAS is thus equivalent to finding node labeling which is unique, and that has as few up-arcs as possible. The most straightforward way of achieving this is by using Depth-First search and labeling the nodes in order they are visited. For the algorithm please consult 3.39.1.

The second is greedy approach based on intuition that nodes with large out-degree need to be at the top of the tree, i.e., their label value would need to be high. Therefore a node with maximal $d_G^+(v)$ is given the label $\delta(v) = |V|$, afterward the node v' with maximal $d_{G-v}^+(v')$ is given the label $\delta(v') = |V| - 1$, etc. Pseudo-code for this procedure can be seen in Algorithm 3.39.2.

The main components of the genetic algorithm proposed by the authors of the paper in question are: representation, crossover operator, mutation operator, fitness function, elite transmission, parents transmission, and termination condition. Detailed explanation of these follow:

1. **Representation**: Permutation of the vertices of G as a representation of a solution for the Feedback Arc Set problem.
2. **Crossover operator**: The algorithm uses two random points, $0 < i < j < n$. Call the elements between V_i and V_j over ordered set S. Delete the elements

---

Algorithm 3.39.1 Simple FAS heuristic [59]

$label \leftarrow 0$
mark each node as unvisited
for all unvisited node v **do**
   $dfs(v)$
end for

function $dfs(v : node)$
mark v as visited
$\delta(v) \leftarrow label + 1$
increment $label$
for all unvisited node $u \in N_G^+(v)$ **do**
   $dfs(u)$
end for
end function

Algorithm 3.39.2 Greedy FAS algorithm [59]

$label \leftarrow |V|$
while $G \neq \emptyset$ **do**
   $v \leftarrow$ the node in G with maximal $d_G^+(v)$
   $\delta(v) \leftarrow label$
   $label \leftarrow label - 1$
   $G \leftarrow G \setminus v$
end while

---

of S from the first parent (leave blank space). Traverse the second parent, and whenever meeting an element of S, put it in the next blank space of first parent. If it is the Kth time meeting an element of S in the second parent, replace it with the Kth element of S. For an example where $i = 3$ and $j = 6$ one can consult Fig. 3.4.

3. **Mutation operator**: Individual $V = (v_1, \ldots, v_n)$ is chosen for mutation. Two indices are first randomly chosen, $\{(i, j) \mid i \geq 1, j \leq n\}$—e.g., $i < j$, components $(v_{i+1}, v_{i+2}, \ldots, v_j)$ are moved to the positions of $(v_i, v_{i+1}, \ldots, v_{j-1})$, respectively, and then move v_i into position of v_j. Mutation probability is changed in different generations so as to increase the efficiency of algorithm. At first, mutation probability is at most 1. If mutation probability in a step is p and a is a small constant, such that $0 < a < 1$, for the next step one has, $p = p - a$.

Fig. 3.4 Crossover operator example (with $i = 3$ and $j = 6$)

FIRST PARENT	SECOND PARENT
4 9 8 6 2 7 3 5 1	2 8 5 4 6 9 1 3 7
4 9 8 2 6 3 7 5 1	6 8 5 4 2 9 1 7 3
FIRST CHILD	SECOND CHILD

4. **Fitness function**: Fitness function is given by the formula $n^2 - x$, where n represents number of nodes of a graph, and x represents the number of return edges determined by the chromosome. Therefore, the less of these return edges, a more quality of a solution one has. Among different functions the authors have tried, the selected one gave best results.
5. **Elite transmission**: The best chromosome so far—is always placed as the first individual of the next generation.
6. **Parents transmission**: If population has m individuals, by roulette wheel, two parents are used, and two children are generated for $m/2$ times. Now the number of individuals have increased to $2m$, and the process is repeated.
7. **Termination condition**: Convergence of the population to a specific percentage.

For the conducted experiment initial population size, initial mutation probability, mutation probability reduction coefficient and convergence percentage were, respectively, as follows: 10, 0.01, 0.001 and 100%. Genetic algorithm was tested on 10 node graphs, while a number of repeated executions was also 10—algorithm was compared to three other algorithms from the literature. Experiment was additionally conducted on 20 and 30 node graphs as well. Genetic algorithm from the paper in question performed better on average, for details one can consult table 1 and figure 3, of the paper in question.

Results were similar for 20 and 30 node graphs, with authors GA and GA from the literature being very similar in a produced solution. When comparison between these two GAs was done on 30 node graph and for 50 runs, new GA proposed in the paper in question again outputted FAS quite similar in size, but average repetition till termination, average number of times reaching suitable result and average runtime were significantly improved, over the literature GA. For additional details one should consult tables and figures near the end of the paper in question.

3.40 Packing Feedback Arc Sets in Reducible Flow Graphs [131]—June 2015

The research presented in this work has presented an $O(n^2 m)$ algorithm (where $n = |V|$ and $m = |A|$) for finding a maximum Feedback Arc Set packing in reducible flow graph (RFG)—for more details on RFG one can consult [71, 72, 89, 127].

The research of the authors was heavily relied on the work from [110, 111], that is on the structural analysis for Reducible Flow Graphs from aforementioned papers. Aside from the algorithmic contribution, the authors of the paper in question also gave proof that every Reducible Flow Graph is feedback arc set Mengerian—for every non-negative integral weight function it satisfies the min-max relation.

The algorithm in question is iterative and in that sense pseudo-polynomial, it needs to be repeated K times in order to acquire maximum FAS packing. It is possible for different stages to return the same FAS, therefore these stages can be merged—by making largest jump on stages, in exchange for efficiency, construction ends in linear time. For a pseudo-code please consult Algorithm 3.40.1.

Algorithm 3.40.1 Maximum FAS packing

Input: An arc-weighted Reducible Flow Graph $G = (V, A; r)$.
Output: A maximum FAS packing $\mathcal{F}$ of G.

construct $N(r, t)$ and define the length function l
$d \leftarrow max_{v \in V_h} d(v)$
$\mathcal{F} \leftarrow \emptyset$

while $d < d(t)$ **do**
    $\mathcal{T} \leftarrow \emptyset, F \leftarrow \emptyset$
    $\alpha \leftarrow d(t)$
    $Augment(\partial^+(U_{d+1}))$
    delete newly added arcs in F and replace remaining arcs by corresponding arcs in G
    $\mathcal{F} \leftarrow \mathcal{F} \cup \{(F, \alpha)\}$
    $Update(l)$
    find $d(v)$ for $v \in V(N)$ with respect to the latest l
    $d \leftarrow max_{v \in V_h} d(v)$
end while
return $\mathcal{F}$

function $Augment(S)$
$\mathcal{T} \leftarrow \mathcal{T} \cup \{S\}$
$F \leftarrow F \cup S$
$\alpha \leftarrow min\{\alpha, min_{(u,w) \in S}\{d(w) - d\}\}$
if $\exists (v', t) \in S$ **then**
  $d \leftarrow max_{(v', t) \in S} d(v)$
  $Augment(\partial^+(U_{d+1}))$
end if
end function

function $Update(l)$
for $T \in \mathcal{T}$ **do**
  for $a \in T$ **do**
    $l(a) \leftarrow l(a) - \alpha$
  end for
end for
end function

In Algorithm 3.40.1, $d(v)$ for $v \in V(N)$ is calculated with respect to the latest length l, repeatedly, with all the single source shortest distances in an acyclic digraph found in $O(n + m)$—where $n = |V|$ and $m = |A|$. For a proof of the $O(n^2 m)$ runtime one should consult theorem 3.7 of the paper in question.

3.41   Algorithms and Kernels for Feedback Set Problems in Generalizations of Tournaments [12]—July 2015

The paper under the title of this section dealt with a number of issues with regards to kernelization, including with a problem of interest to us, namely Feedback Arc Set. The authors have devised polynomial-time kernelization algorithm, on several digraph classes, where given an instance (D, k) of the problem, an equivalent instance (D', k') is returned such that the size of D' and k' is at most $k^{O(1)}$.

Previous results for FAS have been extended, for tournaments, to much larger and well studied classes of digraphs. Polynomial kernel for k-FAS have been obtained on some Φ-decomposable digraphs (round decomposable digraphs), including quasi-transitive digraphs. The authors have designed sub-exponential algorithm for k-FAS (a problem where one asks whether there is a subset of arcs of size at most k such that the digraph obtained after deleting those arcs from digraph is an acyclic digraph) whose running time is $2^{O(\sqrt{k}(\log k)^c)}n^d$—for constants c and d on locally semi-complete (local tournament) digraphs.

A digraph is called semi-complete when there is an arc between every two distinct vertices—a semi-complete digraph without 2-cycles is called a tournament. A digraph D is a local tournament if for every $x \in V(D)$, out-neighborhood and in-neighborhood of x induce semi-complete digraphs (tournaments).

Kernelization is tight if for every instance (x, k) of a problem, the kernel (x', k') is such that $k' = k$ and $\forall h \le k$, (x', h) is a YES instance iff (x, h) is a YES instance. And a kernel $(x'k')$ is immaculate if $x' \subset x$, $k' = k$, and every minimal feedback set of the original graph is also a minimal feedback set for the kernel and vice versa—one should take note that an immaculate kernel is tight, but the reverse need not be the case. A polynomial time decision algorithm implies a tight kernel, but not necessarily an immaculate kernel.

In [17] it has been proved that the problem k-FAS has a kernel with $O(k)$ vertices on tournaments. The authors of the paper being summarized here furthermore state that the problem k-FAS has a tight $O(k)$ kernel on semi-complete multi-digraphs. By following authors material in section 4, Kernels for k-FAS, we can state that k-FAS on totally Φ-decomposable digraphs has an $O(k \cdot f(k))$ kernel—for inductive definition of total decomposition of D one should consult 325-th page of the journal in question. Consequently, the authors have proved corollaries where there is a $O(k^2)$ kernel for k-FAS on quasi-transitive digraphs and on round decomposable digraphs.

The algorithm for k-FAS on locally semi-complete digraphs that the authors have devised is inspired by an algorithm from [4] for (weighted) FAS in tournament, while algorithm from [50] is also used. Considering absence of pseudo-code, and without repeating large parts of the paper, the reader is refereed to section 5 of the paper in question, the authors have presented both randomized and deterministic approach (starting theorem for a general picture would be theorem 5.8), and have used Dynamic Programming as a tool—full algorithm complexity is as follows,

$O(n^3 \log n \cdot 2^{O(\sqrt{k}(\log k)^{O(1)})} + n^2 M(n) \log^2(n))$—where $M(n)$ is the complexity of matrix multiplication [50] and n represents number of vertices.

3.42   An Exact Method for the Minimum Feedback Arc Set Problem [10]—December 2015 (first appeared[4])

An exact method for minimum cardinality Feedback Arc Set problem was the focus of the paper in question. The minimum Set Cover formulation of the minimum Feedback Arc Set problem is appropriate as long as all the simple cycles in G can be enumerated—nevertheless, even sparse graphs can have $\Omega(2^n)$ simple cycles (such graphs appear in practice). For an instance of sparse graphs, the authors have proposed an exact method that enumerates simple cycles in a lazy fashion, and extends an incomplete cycle matrix iteratively. In cases encountered during research, only a tractable number of cycles had to be enumerated until a MFAS is found. Experiments were conducted on computationally challenging sparse graphs that are relevant for industrial applications.

Approach of the authors have been by them termed, an Integer Programming approach with lazy constraint generation. Approach is based upon traditional set covering formulation from [19, 108]. If enumerating all simple cycles of G happens to be tractable (more on simple cycles enumeration in [77]), the integer program with the complete cycle matrix can then be fed to a general-purpose integer programming solver—for example, to Gurobi [67] or to SCIP [1]. The integer program referred can be seen in the minimum set cover formulation from Eq. 3.9.

$$\min_{y} \sum_{j=1}^{m} w_j y_j$$

$$s.t. \sum_{j=1}^{m} a_{ij} y_j \geq 1 \quad \text{for each } i = 1, 2, \ldots, \ell \tag{3.9}$$

$$y_j \text{ is binary}$$

With m representing number of arcs, w_j representing non-negative weights (often integer), y_j is 1 if arc j is in the FAS, and otherwise 0, a_{ij} is 1 if arc j participates in cycle i, and 0 otherwise, and ℓ is the number of simple cycles. The matrix $A = (a_{ij})$ is termed the cycle matrix. This matrix can in practice be substantially reduced in a presolve phase, e.g., by iteratively removing dominating rows and dominated columns of the cycle matrix, and by removing columns that intersect a row with a single nonzero entry.

---

[4] https://www.mat.univie.ac.at/~neum/ms/minimum_feedback_arc_set.pdf.

The method proposed by the authors enumerates simple cycles in a lazy fashion, and extends an incomplete cycle matrix iteratively. In all practical cases the authors encountered, a tractable number of simple cycles had to be enumerated until a minimum FAS is found. For a pseudo-code of the approach one can consult Algorithms 3.42.1 and 3.42.2. Minimum feedback arc set is being found in both terminating cases, and a finite termination is a guarantee—namely, cycle matrix must grow by at least one row in each iteration, and there is only a finite number of simple cycles in the graph. The algorithm first finds FAS with a greedy heuristic. Then tightest simple cycles that contain this feedback arc set is found. After that, the cycle matrix is extend with these simple cycles, they are appended as new constraints.

The authors have also conducted experiments and have presented computational results—for spare graphs the algorithm performed efficiently. During the presolve phase an attempt is made so as to generate an equivalent but simpler graph than the input one. The following procedures were applied: splitting into nontrivial SCCs, then iteratively removing runs and 3-edge bypasses—for more details, one should consult the paper in question (appendix A.3 and figure 4).

Experiments were conducted on Intel(R) Core(TM) i5-4670S CPU at 3.10 GHz, with operating system being Ubuntu 14.04.3 LTS with 3.13.0-86-generic kernel. Integer programming solver Gurobi [67] was called through its API from Python 2.7.11, while the graph library that was used was NetworkX 1.9.1 [68].

After experiments were conducted and results compared it was shown that the median execution time of the proposed method is consistently less than that of the integer programming formulation with triangle inequalities—median execution time increases for both methods as the graph becomes denser. Considering the approach has been devised in order to use it on sparse graphs, the authors have also included, in the experiment, random tournaments in order to examine the worst-case behavior of the proposed method.

The proposed method in such a scenario performs better than the aforementioned integer programming formulation, for $n < 33$ (with n being number of nodes), while in cases where $n \geq 33$, integer programming formulation outperforms the proposed method for random tournaments. A results not surprising with integer programming formulation being specially designed for tournaments. Complete graphs were not an issue for both methods.

Experiments on de Bruijn graphs [44] were conducted as well and the results can be seen in table 2 of the paper in question. Run-time ranged from approx. $1s$, through 20 and 30 seconds, up until $2224s$. Experiments were also conducted for directed graphs of Imase and Itoh [75], results varied but were similar to previous experiment. These two types of digraphs were considered intractable for the methods of integer programming formulation with triangle inequalities and integer programming formulation as minimum set cover, covered in the paper in sections 3.1 and 3.2, respectively. Exploiting fixed-parameter tractability was deemed not an option either, due to the size of the MFAS of these graphs.

Algorithm 3.42.1 Finding minimum FAS—based on integer programming and lazy constraint generation

Input: G, a directed graph with m edges and non-negative edge weights $w_j (j = 1, 2, \ldots, m)$.
Output: A minimum weight FAS.
{P denotes the integer program 3.9 with the complete cycle matrix of G. Optimality is viewed in terms of a relaxed problem.}

let $\hat{y}$ denote the best feasible solution to P found at any point during the search (incumbent solution)

compute a FAS $F^{(0)}$ of G using, for example, minimum set cover heuristic, greedy local heuristics, sorting heuristics, heuristic based on Depth-First search and local search, or heuristics for the closely related Linear Ordering problem (non-exhaustive literature is as follows, [21, 25, 28, 30, 46, 47, 77, 91, 105])

set the solution associated with $F^{(0)}$ as the incumbent $\hat{y}$
set the lower bound $\underline{z}$ and the upper bound $\bar{z}$ on the objective to 0 and $\sum w_j \hat{y}_j$, respectively
let $A^{(i)}$ denote the incomplete cycle matrix in 3.9, giving the relaxed problem $\tilde{P}^{(i)} (i = 1, 2, \ldots)$

call Algorithm 3.42.2 with G, $F^{(0)}$, and an empty cycle matrix to get the first cycle matrix $A^{(1)}$

for $i = 1, 2, \ldots$ **do**
    solve the relaxed problem $\tilde{P}^{(i)}$; results: solution $y^{(i)}$, the associated feedback arc set S and objective value $z^{(i)}$
    {optional: when integer programming solver is invoked on the line just above, $\hat{y}$ can be used as a starting point}
    set the lower bound $\underline{z}$ to $max(\underline{z}, z^{(i)})$
    if $\underline{z} = \bar{z}$ **then**
        return optimal $\hat{y}$
    end if
    let $G^{(i)}$ denote the graph obtained by removing all the edges of S from G
    if $G^{(i)}$ can be topologically sorted **then**
        return optimal $y^{(i)}$ {solution to P}
    end if

    compute a FAS $F^{(i)}$ of $G^{(i)}$ using any of the aforementioned heuristics
    set those components of $y^{(i)}$ to 1 that correspond to an arc in $F^{(i)}$ {$y^{(i)}$ is now a feasible solution to P}
    let $\hat{z}$ be the new objective value at $y^{(i)}$
    if $\hat{z} < \bar{z}$ **then**
        set $\bar{z}$ to $\hat{z}$
        set $\hat{y}$ to $y^{(i)}$
    end if

    call Algorithm 3.42.2 with $G^{(i)}$, $F^{(i)}$, and $A^{(i)}$ to get the extended cycle matrix $A^{(i+1)}$
    {$A^{(i+1)}$ is guaranteed to have at least one additional row compared to $A^{(i)}$}
end for

Algorithm 3.42.2 Extending the cycle matrix—given an arbitrary FAS

Input: G, a directed graph; F, a feedback arc set of G; the incomplete cycle matrix A.
Output: The extended cycle matrix A.

> **for all** $e \in F$ **do**
>> find a shortest path p from the head of e to the tail of e with Breadth-First-Search (BFS) in G
>> **if** such a path p exists **then**
>>> turn the path p into a simple cycle s by adding the arc e to p
>>> add a new row r to the cycle matrix corresponding to s if r is not already in the matrix
>> **end if**
> **end for**

Test graphs used in this research, together with their minimum FAS are available online,[5] with the source of the proposed method also being available at mentioned web-place.

3.43   Monte-Carlo Randomized Algorithm for Minimum Feedback Arc Set [87]—April 2016

Research of the paper in the title has produced, as its main contribution, Monte-Carlo Randomized Algorithm for MFAS, and with this algorithm it has solved the problem of development/execution order of information system subsystems (which uses MFAS as its component). The algorithm solves MFAS in polynomial time with arbitrary probability.

The Monte-Carlo algorithm developed is finding minimum FAS on multi-graph—with multi-graph being a graph which: allows multiple arcs between every pair of nodes, has no loops, and has no arc weights. The main idea for the algorithm is inverse proportionality of uniform arc breaking in a minimal way to the probability of choosing an arc between pairs of nodes connected with minimal number of arcs.

The probability to break all minimal sets of arcs without which a multi-graph would remain acyclic is $\left(\frac{n-2}{n}\right)^{\frac{n}{2}}$. If the algorithm is repeated multiple times this probability is increased. For some number of repeats p_i, probability that we have broken incorrect set of arcs is

$$\left(1 - \left(\frac{n-2}{n}\right)^{\frac{n}{2}}\right)^{p_i} \tag{3.10}$$

---

[5] https://sdopt-tearing.readthedocs.io/en/latest/

Consequently, probability that after p_i number of times one has broken optimal/correct set of arcs is as follows.

$$1 - \left(1 - \left(\frac{n-2}{n}\right)^{\frac{n}{2}}\right)^{p_i} \tag{3.11}$$

MFAS Monte-Carlo Randomized Algorithm was implemented in Python 3.4. The algorithm itself is divided into two functions. One is in charge of finding a solution, and the other is executing the former some specified number of times. Pseudo-code for these can be seen in Algorithms 3.43.1 and 3.43.2—with $TBSL$ being an actual number of sets of arcs, of input graph, which can be found by inspecting neighbors list.

Algorithm 3.43.1 Monte-Carlo randomized algorithm—first function

---

Input: Neighbours list $(V_{\bar{n}}, V_k)$, $V_k \in V$; $V_k \neq V_{\bar{n}}$, number of repeats, and shift. All nodes must be in neighbours list and input arcs must be in groups.

Output: $P(optimum)$, $\sum_{1 \leq u \leq (|V|-1)}^{(u+1) \leq v \leq |V|} w(v, u)$, $v_i (i = 1, \ldots, |V|)$.

    function $repeat(graph, repeat, shift)$
    set pseudo-random number generator seed
    while $repeat$ **do**
        $temp = montecarlo(graph, shift)$
        if better solution found **then**
            $best\_solution = temp$
        end if
        decrease $repeat$ by 1
    end while
    calculate probability $1 - \left(1 - \left(\frac{n-2}{n}\right)^{TBSL}\right)^{p_i}$

    return $best\_solution$ and associated probability
    end function

Input for the aforementioned algorithm is multi-graph. After input is received, the algorithm is uniformly breaking arcs, and in this manner cycles. When acyclic graph is obtained, Topological Sorting algorithm is executed in order to acquire sequence of nodes—this, however, is not essential to the algorithm, the algorithm can therefore be used only for a search for minimum FAS. Complexity of the algorithm is $O(k|V|^3)$. Implementation is based upon data structure known as dictionary.

After experiments and empirical analysis, it has been shown that the algorithm often finds optimum, even when variable *shift* equals 0. The algorithm was finding solutions that were within 0.5 ± 0.8498 feedback arcs from optimum—relatively, the algorithm was finding solutions that were within $1.0973 \pm 1.8937\%$ from optimum. Therefore, Monte-Carlo Randomized Algorithm either finds optimum, or in the worst case a solution that is on average 3% away from optimum. There were

---

Algorithm 3.43.2 Monte-Carlo randomized algorithm—second function

---

function *montecarlo(graph, shift)*
while unvisited arcs exist **do**
    by detecting sets of arcs determine $TBSL$
end while

number of sets of arcs to be broken is, $n\_broken = \frac{TBSL}{2} - shift$
while $n\_broken > 0$ **do**
    uniformly generate (u, v)
    check whether (u, v) belongs to a cycle—by Depth-First-Search
    if generated arc (u, v) does not belong to a cycle **then**
        if number of generated arcs, in a row, is greater than the number of arcs in a graph **then**
            discontinue generation of arcs
            return to, uniformly generate (u, v)
        end if
    end if
    break an arc
    if we have broken last arc in a set of arcs **then**
        decrease n_broken by 1
        store a number of arcs in a set
    end if
end while

calculate $\sum_{(u,v) \in E, u > v} w(u, v)$
find a sequence of nodes $v_i (i = 1, \ldots, |V|)$, with Topological Sort

return $\left(\frac{n-2}{n}\right)^{TBSL}, \sum_{(u,v) \in E, u > v} w(u, v), v_i (i = 1, \ldots, |V|)$
end function

---

nevertheless hard instances, e.g., arc sets close in cardinality, where the algorithm had difficulty finding the right solution.

Empirical running time ranged from a couple of seconds to a couple of hours. Empirical error of the algorithm was determined with the help of Branch and Bound and known optimal solutions for tested digraphs. For a visual example of the algorithm one can consult Fig. 3.5.

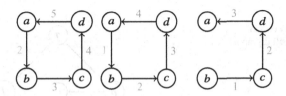

Fig. 3.5 Monte-Carlo Randomized Algorithm example (arc weights represent number of arcs between nodes of a multi-graph), the example assumes perfect uniform arc removal between states

3.44   Optimal Disruption of Complex Networks [134]—May 2016

Here, the authors have dealt with intertwining structure and dynamics in complex networks, as networks relative size indicates cohesion, and it also composes of all the cycles in the network. The authors effort was to find an optimal strategy with which one can remove back-arcs—thus deactivating directed interactions. In this way strongly connected components could be fragmented into tree structures with no effect from feedback mechanism.

The problem researched in the paper, optimal network disruption problem, was mapped into the minimal Feedback Arc Set problem. The problem was solved with statistical physical methods from Spin Glass theory [100]. In such a way a simple numerical method for extraction of suboptimal disruption arc sets was found. Experiments have showed that this way of tackling MFAS proved a better angle of attack than a local heuristic method, and a simulated annealing method—tested in random and real networks.

The par of strongly connected components and loops or cycles have been shown—for a visual example one can consult Fig. 3.6. It is clear how emphasized arc, together with its corresponding vertices, results in the removal of all the cycles. The destruction of all the SCCs in a directed system corresponds to the removal of all the cycles in that same system. Therefore two procedures are obvious, removing of vertices or removing of arcs, with removing of arcs fundamentally dealing with MFAS—and this arc removal is the approach the authors have taken, since the removal of arcs is a more controlled way of local perturbation of network structure.

Based on the height relation, where $h_i > h_j$, with each node being assigned positive integer $h_i \in [0, H - 1]$ as its height, while H is the maximal height chosen for directed graph—this relation accounts for direction of each arc—all the arcs violating the height relation constitute a FAS. The approach is therefore, given a finite maximal height H, to assign the heights on nodes with as many satisfied height relation $h_i > h_j$ for any arc (i, j) as possible—with the finite H providing an upper bound for the size of MFAS.

With this height relation, a spin glass model for MFAS was defined. As MFAS is an optimization problem, the weight sum of FAS has to be minimized, $W(\underline{h}) \equiv$

Fig. 3.6 Optimal network disruption of a small digraph

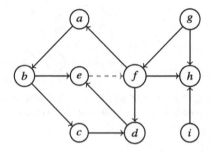

$\sum_{(i,j)\in A} w_{ij}[1 - \Theta(h_i - h_j)]$ for a given H, with reweighing parameter (inverse temperature) having a role. In the framework of Cavity Method of the Spin Glass theory the authors have derived belief propagation algorithm, while dealing with other thermodynamic quantities of the spin glass model, such as the free energy density and the entropy density.

This message-passing algorithm is the belief propagation-guided decimation (BPD). For a given graph instance, BPD follows an iterative procedure consisting of three consecutive steps: graph simplification, message updating, and arc decimation.

1. **Graph Simplification** Using Tarjan's method [126] to exact all SCCs from the original graph or the residual graph—making sure that the cavity messages are only defined and considered on those arcs in the SCCs.
2. **Message Updating** Iterating messages following a randomized sequence of arcs until the iterations converge or reach a maximal number of times.
3. **Arc Decimation** Calculate the marginal probability on each arc in the residual SCCs, and remove those arcs with a given size (for example, 0.5% of the number of remained arcs) with the largest marginals.

Algorithm steps are repeated until there is no SCC. All the removed arcs, by decimation, constitute a suboptimal FAS solution. In order to compare the results obtained by statistical physics, the authors have compared their own approach with two other—namely a local heuristic [104] and a Simulated Annealing [58], which intrinsically involve no notation of height.

All three algorithms were tested on directed random networks with uniform weight on each arc. Testing was conducted on instances of directed Erdős-Rényi random graphs with Poissonian degree distributions and directed regular random graphs with a uniform total degree for each node. Generated directed scale-free networks were also a part of empiricism. For more details one should consult figure 2 of the paper in question—BPD method achieves best results.

Algorithms were also tested on directed real networks. Among 19 datasets, excluding one real network, BPD achieved the smallest FAS size to disrupt the network, especially on networks with moderate size in terms of nodes. Randomized counterparts of the graphs, with connection topology maintained but direction for each arc randomized, have also been considered. While compared with original networks, SCC fraction and FAS size for each type of real network showed a rather similar pattern. Approximated time complexity of the BPD arc decimation is $O(HM \log M)$, with M being number of arcs.

3.45 Efficient Computation of Feedback Arc Set at Web-Scale [123]—November 2016

This paper presents research where several approximations have been investigated, for computing a minimum feedback arc set, with the goal of comparing quality of the solutions and running times. Motivation for this work was application in Social Network Analysis (SNA)—such as misinformation removal and label propagation. In order to achieve web-scale applicability the authors have optimized two approaches, namely greedy and randomized. In this way a balance has been struck between feedback arc set size and running time. Empirical analysis has also been conducted, with a broad selection of large online networks (including Twitter, LiveJournal, and Clueweb12). The algorithms considered were also extended to probabilistic case where arcs were realized with some fixed probability—experimental comparison was conducted here as well.

There are various algorithms presented in the paper in question, but we will focus only on those that are in line with the goal of the research and state of the art in mind. Algorithm that offered best performance, solution wise, was GreedyFAS [47]. A direct implementation, as per the authors of the paper in question, of this algorithm runs in $O(n^2)$ time—which is impractical for large social and web networks (this algorithm pseudo-code can be seen in Sect. 3.13 of the book).

In order to bring this complexity down the authors have made two versions of the algorithm. In the first (**dllFAS**), a custom doubly linked list was implemented for the bins, in order to directly manipulate the list nodes and alleviate a bottleneck suffered by generic (library) lists. In the second (**ArrayFAS**), doubly linked lists are completely excluded from the algorithm, and have been replaced with three flat arrays that mimic the behavior of the lists. Complexity of the algorithm, implemented by the authors of the paper in question, is stated as $O(m + n)$, both in time and space. The approximation guarantee is $\frac{1}{2}|E| - \frac{1}{6}|V|$, but an observation has been made in the experiments, the size of the FAS produced was drastically smaller than the size suggested by the worst-case bound.

For a visual representation of the step where node 4 is processed, please consult Fig. 3.7. $\delta(u) = \delta^+(u) - \delta^-(u)$, i.e., a difference between out-degree and in-degree. In each iteration, the algorithm removes vertices from G that are sources/sinks, followed by a vertex u for which $\delta(u)$ is currently a maximum. The intuition behind this approach is to greedily move all the "sink-like" vertices to the right side of the ordering, and all the "source-like" vertices to the left side, in an attempt to minimize the number of arcs oriented from right to left—pseudo-code can be seen

Fig. 3.7 Processing vertex 4 during the execution of GreedyFAS

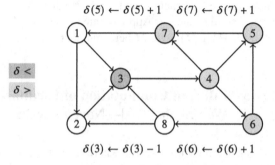

in Algorithm 3.13.1. If this procedure runs through to its completion, FAS of size 1 will be extracted, which corresponds to arc $(3, 4)$ from Fig. 3.7.

Another algorithm for which the authors of the paper in question offered optimization was SortFAS [21], paper of which has been presented in Sect. 3.37 of the book. A direct implementation of this algorithm, as per the authors of the paper in question, runs in $O(n^3)$. With optimization included, this has been reduced to $O(n^2)$—under the assumption that arc membership can be tested in constant time. With adjacency list implementation, run-time complexity becomes $O(n^2 \log(d_{max}))$—where d_{max} is the maximum vertex degree in G.

This algorithm is in its essence sorting by Insertion for the Linear Arrangement problem, and is monotone. The optimal position is defined as the position with the least number of backward arcs induced by v_i, where in the case of a tie the leftmost position is taken. As per insertion sort mechanism, only the arcs between v_i and the first $i - 1$ vertices are relevant in the i-th iteration. In order to decrease run-time, an approach was devised through which it is possible in a single pass to identify best location of v_i, over the $i - 1$ possible locations.

SortFAS—Single Pass Location Identification One begins by initializing a counter variable to zero. Then, for each possible location, j, determine if there is an arc from v_i to v_j and from v_j to v_i. Increment, or decrement, the counter variable if the arc from v_i to v_j, or v_j to v_i, is present, respectively. This process assumes that v_i and v_j will swap locations, and thus keeps track via the counter variable whether or not any current arcs between v_i and v_j would switch direction by inserting v_i at position j. Pseudo-code for the entire algorithm can be seen in Algorithm 3.45.1, while visual representation of the iteration can be seen in Fig. 3.8. In this iteration, node 8 in its original position induces two backward arcs, shown in red, namely $(8, 3)$ and $(8, 2)$. After this particular node is placed at the position of node 2, with node 2 and subsequent nodes moving to the right, this is reduced to one backward arc, namely $(6, 8)$.

The authors have also made modifications to the randomized algorithm BergerShorFAS [16], paper of which has been presented in Sect. 3.9 of this book. This algorithm computes, as the authors have stated, reasonably small FAS while running in $O(m + n)$ time.

The essence of this algorithm is (for MAS was it originally developed) processing vertices of G by selection of either incoming or outgoing arcs, resulting with acyclic graph. While by selecting, at each step, the set of arcs of bigger size, ensures that resulting acyclic graph has a large number of edges. Therefore, $G' = (V, E')$

Fig. 3.8 Iteration 8 of SortFAS algorithm

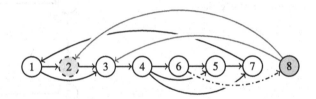

Algorithm 3.45.1 SortFAS

Input: Linear arrangement A.

for all vertices v in A **do**
    $val \leftarrow 0, min \leftarrow 0, loc \leftarrow$ position of v
    for all positions j from $loc - 1$ down to 0 **do**
        $w \leftarrow$ vertex at position j
        if arc (v, w) exists **then**
            $val \leftarrow val - 1$
        else if arc (w, v) exists **then**
            $val \leftarrow val + 1$
        end if
        if $val \leq min$ **then**
            $min \leftarrow val, loc \leftarrow j$
        end if
    end for
    insert v at position loc
end for

---

is returned as an acyclic subgraph, by which $E \setminus E'$ is a feedback arc set. This approach has acyclic subgraph containing at least $\left(\frac{1}{2} + \Omega \left(\frac{1}{\sqrt{d_{max}}} \right) \right) |E|$ arcs. During experimentation it has been detected that BergerShorFAS far outperformed the worst-case bound provided. For a pseudo-code of this procedure please consult Algorithm 3.45.2.

Visualization for the algorithm, with an initial order of $[3, 6, 4, 8, 7, 1, 2, 5]$, is presented in Fig. 3.9. Vertex 3 has three incoming arcs, and one outgoing arc. Incoming arcs are therefore added into E', after which arcs of this node are removed from G. This process is repeated, until finally all vertices are processed—at this stage $E \setminus E'$ constitutes a feedback arc set. In an instance from Fig. 3.9, FAS is of size 3, and is represented by the set $\{(3, 4), (4, 6), (5, 7)\}$.

In order to reduce memory footprint, status whether vertices are deleted or present is being kept via a bit set—the same is not done for arcs, as it is being inferred from the information known about vertices. In this way, only a single auxiliary data structure of size $O(n)$ is needed, since nothing is kept for arcs.

Another algorithm that was being considered was KwikSortFAS [3], presented in Sect. 3.29 of the book (3-approximation algorithm for tournaments)—the same

Fig. 3.9 Processing vertex 3 during execution of the BergerShorFAS algorithm

$E' \leftarrow E' \cup \{(1, 3), (2, 3), (8, 3)\}$

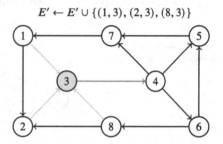

Algorithm 3.45.2 BergerShorFAS

Input: Directed graph $G = (V, E)$.
Output: A feedback arc set for G.

fix an arbitrary permutation P of the vertices of G
$F \leftarrow 0$
for all vertices v processed in order based on P **do**
  if $inDegree(v) > outDegree(v)$ **then**
    $F \leftarrow F \cup \{(v, u) : u \in G.succ(v)\}$
  else
    $F \leftarrow F \cup \{(u, v) : u \in G.pred(v)\}$
  end if
  $E \leftarrow E \setminus (\{(v, u) : u \in G.succ(v)\} \cup \{(u, v) : u \in G.pred(v)\})$
end for
return F

---

algorithm was extended in [21] to work as a heuristic algorithm for general directed graphs, this work is presented in Sect. 3.37 of the book. Motivation for KwikSortFAS was classical sorting algorithm Quicksort. The algorithm uses the 3-way partition variant of Quicksort, due to it being highly adaptive in the case of sorting with many equal keys. Two vertices are considered equal if they do not have arcs connecting them. The authors implementation follows optimization from [120], which uses $O(\log n)$ of additional space.

Given a starting linear arrangement, vertices are moved to the left or to the right, relative to a random pivot element, based on whether there is an arc to or from the pivot—the algorithm then proceeds recursively. Pivot and vertices equal to it (no arc from or to the pivot) are placed in the middle, and then recursion is made on the left, middle, and right subsets. Even though vertices in the middle are equal, they could have arcs between them, and may not be equal to each other, therefore middle needs to be recursively processed (in case of ties, order is left unaltered). Pseudo-code for this algorithm can be seen in Algorithm 3.45.3, and a visual initial recursive step is presented in Fig. 3.10. This initial ordering has node 4 randomly selected as the pivot. Then other vertices are partitioned accordingly to the left, middle, and right. After which one must recourse on each partition. In the end, a FAS can be extracted from final sorted order.

The run-time complexity of this algorithm is $O(n \log n)$, and assumes that arc membership can be tested in constant time (when the graph is represented with an adjacency matrix). The algorithm of the paper in question used an adjacency list, and therefore a search of the list (which is kept in sorted order) needs to be executed

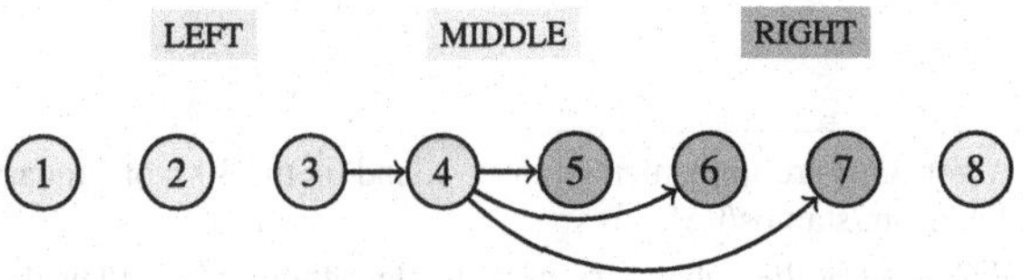

Fig. 3.10 Initial recursive step of KwikSortFAS

Algorithm 3.45.3 KwikSortFAS

Input: Linear arrangement A, vertex lo, vertex hi.

if $lo < hi$ **then**
    $lt \leftarrow lo, gt \leftarrow hi, i \leftarrow lo$
    $p \leftarrow$ random pivot in range $[lo, hi]$
    while $i \leq gt$ **do**
        if arc (i, p) exists **then**
            $swap(lt, i)$
            $lt \leftarrow lt + 1, i \leftarrow i + 1$
        else if arc (p, i) exists **then**
            $swap(i, gt)$
            $gt \leftarrow gt - 1$
        else
            $i \leftarrow i + 1$
        end if
    end while
    $KwikSortFAS(A, lo, lt - 1)$
    if at least one swap was made **then**
        $KwikSortFAS(A, lt, gt)$
    end if
    $KwikSortFAS(A, gt + 1, hi)$
end if

---

so as to test for the presence of an arc. With an adjacency list representation, the runtime complexity is $O(n \log n \log(d_{max}))$. It should be noted, since KwikSortFAS is randomized, each run may yield a different result.

Aside from mentioned algorithms, there were also other, perhaps of a lesser note, which were discussed, these algorithms were: GreedyAbsFAS, SimpleFAS, dfsFAS, InsertionSortFAS, and SiftFAS. All of these can as well be found in section 2, Algorithms, of the paper in question, if one wants inform himself about these.

Experiments that were conducted by the authors[6] were on a machine with dual 6 core 2.10 GHz Intel Xeon CPU, with 32 GB of RAM and running on Ubuntu 14.04.2. Datasets statistic can be seen in the paper in question, in table 1, while the algorithms were summarized in table 2. The measures of effectiveness were FAS size, defined as the number of arcs in a FAS output by a particular algorithm, and the algorithm efficiency, measured in running time—the goal was to keep both parameters as low as possible. Datasets ranged from tens of thousands of nodes and arcs, all the way to web-scale and millions of nodes and billions of arcs—datasets were obtained from Laboratory of Web Algorithmics.[7]

After experimentation the following were the results. It has been shown that GreedyFAS and BergerShorFAS approaches benefit when there are many nodes

---

[6] Algorithms are implemented in Java, and on the 30th of January 2022 were available at: https://github.com/stamps/FAS.

[7] Web page of the Laboratory of Web Algorithmics: http://law.di.unimi.it/datasets.php.

that are sources and sinks, present in G. When sorting algorithms are looked upon, difficulty in applying sorting techniques to the FAS problem is a lack of transitivity which sorting algorithms are designed to exploit, and which exists in traditional sorting problems where there is a total ordering on the data (real-world networks are far from exhibiting a total ordering because of their sparsity). For example, in KwikSortFAS, order of equal vertices is left unaltered, therefore in sparse graphs, where many vertices will probably be equal (in each iteration), this will most likely lead to poor performance—since large subsections of the ordering will not be modified in a meaningful way.

For small datasets, around tens of thousands of nodes and arcs, sorting algorithm were competitive in the output FAS, nevertheless as the size of the graph on which experiments were conducted increased, GreedyFAS and BergerShorFAS showed their superiority, with GreedyFAS being the winner. The authors have also investigated the effect of power-law degree distribution and the small-world phenomena by constructing synthetic networks. While small-world phenomena property of a graph did cause linear decrease in the size of the FAS by both GreedyFAS and BergerShorFAS, power-law degree distribution property of a graph did not cause any meaningful difference.

Investigation into the algorithms and input graphs, in a probabilistic case, has also been performed. Two algorithms were in a focus, probabilistic GreedyFAS, and probabilistic BergerShorFAS. GreedyFAS has seen an adaptation in the form of an approximate δ-class, with BergerShorFAS following a natural extension by altering its decision function for updating set F, probabilistic FAS, to incorporate the probabilities on the arcs. Experimentation has measured the same points, and with the same goal—medium datasets (hundreds of thousands of nodes and millions of arcs) and a subset of the large datasets (millions of nodes and millions to billions of arcs) were considered. Observation has been made that both GreedyFAS and BergerShorFAS algorithms see an improvement, in terms of the expected FAS size, for several datasets compared to the unweighted case—the additional information leads to the algorithms choosing FASs that contain low probability arcs, with the result being a smaller expected value.

Considering the aim of the research was optimization and web-scale application, scalability is an important factor. For the algorithms of $O(n^2)$ observed approximate maximum scalability was $300K$ arcs. For the algorithms of $O(n \log n)$ this was $3.5M$ arcs. And for the algorithms of $O(m + n)$ it as $50B$ arcs. Approximate FAS sizes of 3-20%, 23-40% and 11-17% were achieved—for the best algorithms in each run-time category, $O(m+n)$, $O(n \log n)$, and $O(n^2)$, respectively. GreedyFAS and BergerShorFAS provided the best balance between scalability and quality, with GreedyFAS being the algorithm that always produces either the smallest, or a very close second smallest, FAS size—while being a fast algorithm in general. GreedyFAS (array) implementation was particularly good performer and suited for the biggest dataset considered (clueweb12[8]), with more than 42 billion arcs.

Fig. 3.11 Digraph
D-Segmentation problem
illustration (source digraph
for segmentation in Fig. 3.12)

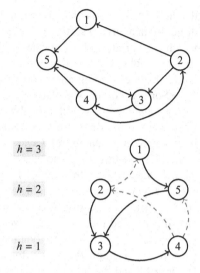

Fig. 3.12 Digraph
D-Segmentation problem
illustration (segmentation for
digraph in Fig. 3.11)

$h = 3$

$h = 2$

$h = 1$

3.46 Optimal Segmentation of Directed Graph and the Minimum Number of Feedback Arcs [132]—August 2017

The paper deals with FAS, that is MFAS, by considering a generalized task of dividing the digraph into D layers of equal size—this problem involving FAS is called D-Segmentation problem. The problem was solved by the Replica-Symmetric (RS) Mean Field Theory and belief propagation heuristic algorithms. The minimum feedback arc density of a given random digraph ensemble is then obtained by extrapolating the theoretical results to the limit of large D. Specifically concerning the problem of minimum FAS, a Divide-and-Conquer algorithm was devised—the algorithm is giving good performance and is contextually efficient.

The authors of the research have observed the problem of FAS as a statistical physical system, through the Optimal Segmentation problem, in which one needs to evenly distribute N vertices of a digraph into D layers, while under the constraint where total number of arcs pointing from lower layers to higher layers needs to be minimized. For a graphical representation one can consult Fig. 3.11 for a source digraph, and Fig. 3.12 for segmentation example. Segmentation example is partitioned into $D = 3$ layers with 3 feedback arcs (dashed red arcs). Each vertex i has a state derived from height h, $1 \leq h \leq D$, with $h_i \in \{1, 2, \ldots, N\}$. Therefore graph in Fig. 3.12 has the following configuration, $h_3 = h_4 = 1$, $h_2 = h_5 = 2$, $h_1 = 3$.

---

[8] Dataset web-page accessed on 31th of January 2022 at: https://law.di.unimi.it/webdata/clueweb12/.

By applying RS Mean Field theory to three different types of random digraphs, for each digraph type, the authors have determined the minimum fraction of feedback arcs for the D-Segmentation problem. These three types of digraphs were: balanced regular random digraph [65] (each vertex has α incoming arcs and α outgoing arcs, with α being an integer), regular random digraph (each vertex is attached the same number of arcs, the direction being chosen randomly), Erdős-Rényi digraph ($M = \alpha N$, direction of each newly added arc ($\in M$) is chosen randomly, two end points of this arc are also chosen uniformly at random from all the N vertices).

For an example of this predictive solution one can consult figure 3 of the paper in question, and its corresponding chapter text. The same theoretical method has been applied to MFAS, for the same types of digraphs and for specific arc densities—for more details one should consult table 1 of the paper in question (depending on α and type of digraph, the predictions ranged from a few percentiles to a few dozen percentiles).

In order to show possible applicability of RS Mean Field theory, the authors have described how a heuristic with Belief Propagation-Guided Decimation can be an efficient solver for the D-Segmentation problem and the minimum FAS problem. This algorithm can additionally be improved by introducing guided Reinforcement Learning. The algorithm with learning mechanism was faster than the one without, and it has also been shown that this algorithm was comparable and was able to compete with well known approach, namely Simulated Annealing (SA), but if SA uses only a single run, while the reinforcement learning algorithm uses multiple recursive calls in order to achieve its result, explained bellow.

Considering that after upward feedback arcs are deleted from the graph, those connecting different layers, directed cycles may still exist within each single layer, the authors have suggested nesting as the solution, that is recursive application of the aforementioned reinforcement learning algorithm. After this nested reinforcement learning process is finished, and all the upward arcs are deleted, the remaining digraph will be free of any directed cycles—while the set formed by all the upward arcs then must be a feedback arc set for the input digraph.

Recursive reinforcement learning algorithm preformed favorably, in terms of a running time, as compared to SA implementation used in the research experiments—in a mentioned instance, 4.4 h compared to 63.1 h, on a computer with 2.5 GHz frequency. As the authors have not given pseudo-code, and without repeating large parts of the paper in question here, the reader is referred to the paper itself (section, Algorithmic Applications, is of interest).

3.47   Quantum Speedups for Exponential Time Dynamic Programming Algorithms [6]—January 2019

As the title of this paper immediately reveals, the authors have in this paper presented their research on quantum algorithms for NP-Complete problems, where best classical algorithms present exponential time application of Dynamic Programming.

In order to do this, the authors have introduced the path in the Hypercube[9] problem,[10] that models many of these dynamic programming algorithms. Here, a question is asked, whether there is a path from 0^n to 1^n in a given subgraph of the Boolean hypercube, where the edges are all directed from smaller to larger by having as the criteria Hamming weight (the Hamming distance between two sequences of symbols is the number of positions in which the symbols are different [129]).

As a general approach, quantum algorithm that solves the path in the hypercube in time $O^*(1.817^n)$ has been given—approach combines Grover's search[11] with computing a partial dynamic programming table. With some similar ideas the $O^*(1.728^n)$ quantum algorithm was devised so as to solve Traveling Salesman (TS) problem and Minimum Set Cover (MSC) problem, as well as the problem that is of interest to us, namely FAS.

There is a subclass of problems, where solution to a problem that involves set S of size n can be calculated by considering all partitions of S into two sets, of size k and $n - k$, and then choosing the best option, for any fixed positive k. TSP admits such a decomposition, since sub-paths of an optimal path should also be optimal. By combining with the algorithm from [66], the quantum algorithm for TSP was obtained, with $O^*(1.728^n)$ running time—this algorithm was adapted in order to serve MSC and FAS (classical best, $O^*(2^n)$ [20]).

The main idea for devised quantum algorithms is as follows. Precompute solutions for a part of the subsets using Dynamic Programming. After which use Grover's search on the rest of the subsets to find the answer one is looking for—the algorithms return correct answer with probability at least $\frac{2}{3}$, while they require exponential space since they need to store a partial dynamic programming table. Pseudo-code for the TSP quantum algorithm can be seen in Algorithm 3.47.1.

Graph $G = (V, E, w)$, where $|V| = n$, $E \subseteq V^2$, and $w : E \rightarrow \mathbb{N}$ are the edge weights—if $\{u, v\} \notin E$ then $w(u, v) = \infty$. $f((S, u, v) \rightarrow \mathbb{N})$ represents length of the shortest path in the graph, induced by S that starts in u, ends in v, and visits all vertices in S exactly once, where $\{S \subseteq V, u, v \in S\}$. With $N(u)$ being the set of neighbors of u in G, and $\alpha \in (0, 1/2]$ being a parameter of the algorithm.

Algorithm 3.47.1 is of another importance, since it is possible to extend it for the problem of FAS. Consider any partition of V into two sets, A and $V \setminus A$ (of size k

---

[9] n-dimensional counterpart of a cube, or a square: https://mathworld.wolfram.com/HypercubeGraph.html.

[10] Finding a specific path alongside edges of a hypercube [81].

[11] A quantum unstructured search algorithm that finds desired item in $O(\sqrt{N})$ steps, with N being number of items to choose from [62].

Algorithm 3.47.1 Quantum algorithm for TS

HypercubePath(subgraph G of Q_n): whether 0^n and 1^n are connected by a directed path.
TravelingSalesman(graph G, edge weights w): length of the shortest Hamiltonian cycle.

1. **Calculate** the values of $f(S, u, v)$ for all $|S| \leq (1 - \alpha)n/4$ classically **using dynamic programming,**

$$\min_{\substack{t \in N(u) \cap S \\ t \neq v}} \{w(u, t) + f(S \setminus \{u\}, t, v)\} \mid f(\{v\}, v, v) = 0,$$

and **store** them **in memory.**

2. **Run quantum minimum finding** over all subsets $S \subset V$ such that $|S| = n/2$ to find the answer,

$$\min_{\substack{S \subset V \\ |S| = n/2}} \min_{\substack{u, v \in S \\ u \neq v}} \{f(S, u, v) + f((V \setminus S) \cup \{u, v\}, v, u)\}.$$

To calculate $f(S, u, v)$ for $|S| = n/2$, **run quantum minimum finding,**

$$\min_{\substack{X \subset S, |X| = k \\ u \in X, v \notin X}} \min_{\substack{t \in X \\ t \neq u}} \{f(X, u, t) + f((S \setminus X) \cup \{t\}, t, v)\},$$

with $k = n/4$. To calculate $f(S, u, v)$ for $|S| = n/4$, **run quantum minimum finding** for previous equation, with $k = \alpha n/4$. For any S such that $|S| = \alpha n/4$ or $|S| = (1 - \alpha)n/4$, we **know** $f(S, u, v)$ **from the classical preprocessing.**

---

and $n - k$, respectively). It is of requirement that in the acyclic graph G' there are no edges from $V \setminus A$ to A, i.e., A is before of $V \setminus A$ in the topological order of G'—minimum FASs for both of these can be computed independently. If $f(S)$ is the size of the minimum FAS of G induced on S, then computation can be made by the following recurrence:

$$f(S) = \min_{\substack{A \subseteq S \\ |A| = k}} \{f(A) + f(S \setminus A) + |\{(b, a) \in E \mid a \in A, b \in S \setminus A\}|\} \qquad (3.12)$$

This recurrence can be used in the quantum algorithm for TS, by replacing the formula in the algorithm, used for repeated run for quantum minimum finding, with the one from Eq. 3.12. This gives $O^*(1.728^n)$ quantum algorithm for directed FAS, which represents an improvement over the previous best-known classical algorithm, with application of Dynamic Programming, which runs in $O^*(2^n)$ time.

3.48　Ant Inspired Monte-Carlo Algorithm for Minimum Feedback Arc Set [88]—May 2019

The paper that presents the research is a continuation of the research presented in [87], with the addition here being learning mechanism. With this new and improved ant inspired Monte-Carlo (AIMC) algorithm running time was on average improved by 20%, with convergence being 511% faster, in terms of median, and 158% in terms of arithmetic mean. At the same time, the ability of the original MC of finding solution with arbitrary probability was maintained in the new and improved iteration.

The main mechanism through which improvements over the original MC was achieved was implemented learning technique inspired by Ant colony Optimization (ACO)—for an example of ACO one can consult [43]. The ant inspired MC, however, does not follow standard ACO framework. In addition, Dynamic Programming was incorporated, and thus execution time reduced. Complete procedure can be seen in Algorithms 3.48.1, 3.48.2 and 3.48.3.

Algorithm 3.48.1 AIMC randomized algorithm for MFAS—function PLEARN

---

function $PLEARN(graph, r)$
for all i **in** range(r) **do**
　$REPEAT(graph, repeat, shift, step)$
end for
$P = 1 - \left(1 - \left(\frac{1}{2} - \frac{1}{n}\right)\right)^r$ {n is the number of nodes in the $graph$}
end $PLEARN$

Algorithm 3.48.2 AIMC randomized algorithm for MFAS—function REPEAT

---

function $REPEAT(graph, repeat, shift, step)$
set the seed of pseudo-random number generator
while $repeat$ **do**
　$temp = MONTECARLO(graph, shift)$
　if $temp < best\_solution$ **then**
　　$best\_solution = temp$
　　if this is the $step$ for multigraph updating **then**
　　　for all arcs **in** $best\_solution$ **do**
　　　　update multi-graph used for learning
　　　end for
　　end if
　end if
　$repeat = repeat - 1$
end while
end $REPEAT$

Algorithm 3.48.3 AIMC randomized algorithm for MFAS—function MONTECARLO

---

function $MONTECARLO(graph, shift)$
$sba = 0$ {total number of broken arcs}
if function is executed for the first time **then**
    while unvisited arcs exist **do**
        determine number of sets of arcs (NSA)
    end while
    $no\_broken = \lfloor \frac{NSA}{2} - shift \rfloor$
    create multi-graph copy (mgc)
end if
$k = 0$ {no. of gen. but not broken arcs—in a row}
while $n\_broken > 0$ **do**
    uniformly generate (u, v)—from mgc
    if (u, v) does not belong to cycle **then**
        $k++$
        if $k >$ number of arcs in $graph$ **then**
            break
        end if
        continue
    end if
    if not known if arc belongs to cycle **then**
        if DFS (u, v) **then**
            store: (u, v) is in cycle
        else
            store: (u, v) is not in cycle
            $k++$
            continue
        end if
    end if
    break an arc from $graph$
    if we have broken last arc in set of arcs **then**
        $n\_broken = n\_broken - 1$
        $sba+ = |set\ of\ broken\ arcs|$
        reset stored information about arcs in cycles
    end if
end while
reset stored information about arcs not in any cycle
$t =$ topologically sorted sequence of nodes
return $\sum_{(u,v)\in A, u>v} W(u, v), t, sba$
end $MONTECARLO$

Learning mechanism has been implemented greedily, so as to push the algorithm in a right direction as soon as possible, and therefore increase convergence. Even though learning mechanism has been implemented into the novel Monte-Carlo algorithm, the ability of the original algorithm to output probability that the optimal solution has been achieved is preserved. Probability of breaking (via uniformly selecting arcs) all cycles in a multi-graph, between pairs of nodes connected with minimal number of arcs, is at least $\frac{1}{2} - \frac{1}{n}$ for the first s times (and arbitrarily grows for every other s times). Therefore probability to optimally break sets of arcs when

$s = 1$ and $r \to \infty$, where r represents number of times s iterations were repeated, is

$$P(optimum, r) = 1 - \left(1 - \left(\frac{1}{2} - \frac{1}{n}\right)\right)^r \tag{3.13}$$

By changing the values of s and r it is possible to turn pure randomized algorithm into one with learning mechanism, and vice versa—without losing probability that optimal solution has been achieved.

Algorithm complexity, for this improved version of ant inspired MC, has changed from what it was in the previous MC algorithm and now amounts to $O(r \cdot s \cdot |V|^3)$, with $r, s = 1 \ldots \infty$.

Experiments for both dynamic programming (space-time trade-off) and for learning mechanism has been conducted. MC randomized algorithm with space-time trade-off, but without learning mechanism, was tested on digraphs ranging from 20 to 50 nodes, with arcs being between 190 and 1225—the same instances on which original MC algorithm was tested. The novel algorithm, that uses space-time trade-off, performed substantially faster than original MC randomized algorithm (on average 19.8% faster, with deviation from arithmetic mean being 7.07%).

By enabling learning mechanism as well, and experimenting on the complete ant inspired Monte-Carlo algorithm, it was determined that the learning mechanism did not have an effect on the algorithm approximation, but it did have substantial positive impact on algorithm convergence toward a solution. The algorithm converged significantly faster toward optimum than original MC algorithm in all tested cases, and on the average it was 511% faster in terms of median number of iterations (158% in terms of mean). Results have also showed that ant inspired MC algorithm rarely performed so pessimistically where it was not able to find a solution that original algorithm would find—this means that the potential error introduced with the learning mechanism is not high. Chance that ant inspired MC algorithm performs in a less number of iterations, while at the same time being generally more consistent, was greater than that of the original algorithm.

3.49   Tight Localizations of Feedback Sets [70]—December 2021

In the paper in question the authors have proposed a new general heuristic for the directed FAS. Algorithm complexity for the proposed heuristic is $O(|V||E|^4)$—via empirical validation the authors have achieved approximation of $r \leq 2$.

Given multi-digraph $G = (V, E, head, tail)$ an algorithm termed ISO–CUT was formulated, this procedure makes a part of the aforementioned general heuristic. The procedure searches for localized optimal arc cuts $e \in E$, if isolated cycles that satisfy the following exists.

Firstly, if $\epsilon \in \mathcal{F}_E(G, w)$ is a minimum feedback arc set, where $e \in \epsilon$, then $\overrightarrow{F}(e) \subseteq \epsilon$—with G and w being multi-graph and arc weight function ($w : E \rightarrow \mathbb{R}^+$), respectively, $\overrightarrow{F}(e)$ representing parallel arcs of e, and $\mathcal{F}_E(G, w)$ denoting set of solutions for FAS problem.

Secondly, if $I_e \neq \emptyset$ and $\delta = mincut(head(e), tail(e), I_e, w_{|I_e}) \in \mathcal{P}(E)$ is a minimum–s–t–cut, where source $s = head(e)$ and target $t = tail(e)$ with respect to I_e and $w_{|I_e}$, such that

$$\Omega_{G,w}(\delta) \geq \Omega_{G,w}(\overrightarrow{F}(e)) \tag{3.14}$$

then there is $\epsilon \in \mathcal{F}_E(G, w)$ with $\overrightarrow{F}(e) \subseteq \epsilon$. Where I_e represents the subgraph induced by all isolated cycles passing through e (if cycle c is isolated, then c intersects with no cycle c' passing not through e or some parallel arc of e), $\mathcal{P}(E)$ is a power set of E, and Ω represents, for a given G and w, and for a certain set of arcs, sum $\sum_{e \in \epsilon} w(e)$ (which is to be minimized for a problem, for weighted FAS $\epsilon \in \mathcal{P}(E)$ where $G \setminus \epsilon$ is acyclic).

If such an arc $e \in E$ is located, it is stored in a list ϵ, $G = G \setminus e$ is considered, and the search is continued—until either the resulting graph G is acyclic or no desired arc can be localized. The stored arcs ϵ are therefore optimal sub-solution for the FAS problem on G. For proof one should consult theorem 2.5 and appendix A of the paper in question, pseudo-code of the procedure can be seen in Algorithm 3.49.1.

Algorithm 3.49.1 ISO-CUT

Input: G, w—w is an arc weight.
Output: (G, ϵ).

$\epsilon = \emptyset, iso = 1$
while G is not acyclic {check by topological sorting in $O(|E|)$ time} **and** $iso = 1$ **do**
  $iso = 0$
  for $e \in E$ **do**
    compute I_e and $\delta = mincut(head(e), tail(e), I_e, w_{|I_e})$

    {2nd condition in the selection is redundant if G is a digraph with $w \equiv 1$}
    if $I_e \neq \emptyset$ **and** $\Omega_{G,w}(\overrightarrow{F}(e)) \leq \Omega_{G,w}(\delta)$ **then**
      $G = G \setminus e$
      $\epsilon = \epsilon \cup \overrightarrow{F}(e)$
      $iso = 1$
    end if
  end for
end while
return (G, ϵ)

Algorithm 3.49.1 needs $O(|E|^3 + |V||E|^2)$ run-time in order to return an optimal sub-solution $\epsilon \in E$ and the remaining graph $G \setminus \epsilon$, for the FAS problem in the unweighted instance. To return the same for the weighted instance, run-time of

$O(|V||E|^3)$ is needed. If the resulting $G \setminus \epsilon$ is acyclic, then $\epsilon \in \mathcal{F}_E(G, w)$ is a minimum FAS. Since loops are isolated cycles, the algorithm ISO-CUT removes every loop from G.

Isolated cycles allow a situation where optimal cuts can be localized, nevertheless they do not need to exist. Therefore procedure in Algorithm 3.49.1 might not return wanted, namely acyclic graph. For such an instance a concept of a good guess was developed, in order to cut G in a pseudo-optimal way, until isolated cycles have come into play, and then ISO-CUT can compute as usual.

If $I_e = \emptyset$ and $G_e \neq \emptyset$, then $\delta = mincut(head(e), tail(e), G_e, w_{|G_e})$ is a minimum–s–t–cut with $\epsilon \in \mathcal{F}_E(G, w)$ being a minimum FAS—while $\epsilon' = \epsilon \cap \mathcal{E}(G_e)$ denotes its restriction to G_e (with $\mathcal{E}(\cdot)$ representing induced set of all arcs). If

$$\Omega_{G,w}(\delta) - \Omega_{G,w}(\overrightarrow{F}(e)) > \Omega(G \setminus \overrightarrow{F}(e), w) - \Omega(G \setminus \epsilon', w) \qquad (3.15)$$

then $\overrightarrow{F}(e) \subseteq \epsilon$—for proof one should consult proposition 3.2 of the paper in question. The right-hand side of Eq. 3.15 is hard to compute, and to estimate. Therefore intuitively, the larger left-hand side is, more likely it is that the inequality in 3.15 holds—this constitutes basic idea for the aforementioned concept of a good guess.

Maximizing the left-hand side of Eq. 3.15 is costly, the authors have therefore restricted considerations to one cycle $c \in O_{el}(G)$ (with $O_{el}(G)$ being the set of all directed elementary cycles) and all arcs $e_1, \ldots, e_n \in \mathcal{E}(c)$ with $\mathcal{G}(c) \subsetneq G_{e_i}$ ($i = 1, \ldots, n$) which cut more cycles than c does—where G_e represents the subgraph induced by all elementary cycles passing through e or parallel arcs $f \in \overrightarrow{F}(e)$ the cycle cover of e. One needs to therefore chose good guess

$$GG(G, w, c) = \underset{e_i, i=1,\ldots,n}{argmax} \left(mincut(head(e_i), tail(e_i)) - \Omega_{G,w}(\overrightarrow{F}(e_i)) \right)$$

$$(3.16)$$

as an arc which has the most expansive minimum s–t–cut for the one to be cut—running time of making this heuristic decision is $O(|c||E||V|)$. This global approach procedure that unites ISO-CUT and good guess 3.16 can be seen in Algorithm 3.49.2.

The procedure TIGHT-CUT described in Algorithm 3.16 outputs solution for FAS $\epsilon \cup \delta \subseteq E$ in $O(|V||E|^3 + |E|^4)$ for the unweighted problem instance. If the instance is weighted, then return of FAS requires $O(|V||E|^4)$ in terms of running time. This same algorithm can be adapted for the problem of FVS (proposition 3.3 of the paper in question).

In order to make devised algorithm TIGHT-CUT more efficient, the authors have created a notion of almost isolated cycles. Graph $G = (V, E, head, tail)$ is a multigraph, while $n \in \mathbb{N}$ and $e \in E$. If there exists a set $\mu \subseteq E$ of n arcs (i.e., $|\mu| = n$ s.t.

Algorithm 3.49.2 TIGHT-CUT

Input: G, w—w is an arc weight.
Output: $(\epsilon \cup \delta)$, $\Omega_{G,w}(\delta)$.

$\epsilon = \emptyset$, $\delta = 0$
for $i = 1, \dots, |E|$ **do**
    $(G, \epsilon') = ISO\text{-}CUT(G, w)$
    $\epsilon = \epsilon \cup \epsilon'$

    {can be checked by topological sorting in $O(|E|)$ time}
    if G is acyclic **then**
        break
    else
        choose $c \in O_{el}(G)$
        $h = GG(G, w, c)$
        $\delta = \delta \cup \{h\}$
        $G = G \setminus h$
    end if
end for
return $(\epsilon \cup \delta)$, $\Omega_{G,w}(\delta)$

$I_e \neq \emptyset$ w.r.t. $G \setminus \mu$), then cycles $c \in O_{el}(I_e)$ are called almost isolated cycles—if $n = 0$ then one has a previous notion.

The idea is that for small enough $n \in \mathbb{N}$ the accuracy of such an algorithm would remain high. If no isolated cycles were found, then N graphs are being generated, $H_i = G \setminus \mu_i$, randomly uniformly deleting arcs—$\mu_i \subseteq E$, $|\mu_i| = n$ $(i = 1, \dots, N)$, $n, N \in \mathbb{N}$. In this way a search for the isolated cycles is being performed, search for arcs f in H_i with $I_f \neq \emptyset$. It is assumed that those arcs that appear in most of the explored graphs represents a good choice for the cut in the original graph. If on the other hand no such arc can be found, GG is used to make the choice—pseudo-code for this procedure is seen in Algorithm 3.49.3.

Relaxed version of the algorithm, namely TIGHT-CUT*, was implemented in C++.[12] Experiments were conducted in a single threaded mode on a machine with Intel CPU (dual Intel(R) Xeon(R) E5-2660 v3 @ 2.60 GHz) and 128 GB of memory. Operating system in use was Ubuntu 16.04.6 LTS, using compiler GCC 9.2.1. Implementations used in the experiments were: an exact integer linear programming–based approach implemented as the feedback_edge_set function from SageMath 8.9 with iterative constraint generation (termed EM), the greedy removal approach from [47] (termed GR—imported from the igraph library [36, 74]), and the relaxed version TIGHT–CUT* (with settings $n = 3$ and $N = 20$).

For a detailed look of the empirical results one should consult section 5.1 (Synthetic Instances) and 5.2 (Real-World Datasets) of the paper in question. For the examples of data for validating the ratio approximation one should consult table 1

---

[12] C++ code, including benchmark datasets, was on 4th of March 2022 available at: https://git. mpi-cbg.de/mosaic/software/math/FaspHeuristic.

Algorithm 3.49.3 TIGHT-CUT*

Input: G, w, n, N—w is an arc weight, while $n, N \in \mathbb{N}$.
Output: $(\epsilon \cup \delta)$, $\Omega_{G,w}(\delta)$.

$\epsilon = \emptyset, \delta = \emptyset$
for $i = 1, \ldots, |E|$ **do**
   $(G, \epsilon') = ISO\text{-}CUT(G, w)$
   $\epsilon = \epsilon \cup \epsilon'$

    {can be checked by topological sorting in $O(|E|)$ time}
   if G is acyclic **then**
      break
   else
      choose $\mu_1, \ldots, \mu_N \subseteq E$ with $|\mu_i| = n$ uniform at random
      $(H_i, \varrho_i) = ISO\text{-}CUT(G \setminus \mu_i)$

      {$\varrho_{i,1}$ is the first arc cut by ISO-CUT}
      $R = \cup_{i=1}^{N} \varrho_{i,1}$
      if $R \neq \emptyset$ **then**
         { f is a good choice in most of the H_i }
         $f = argmax_{e \in R} \left| \{e = \varrho_{i,1}\} 1 \leq i \leq N \right|$
         $\epsilon = \epsilon \cup \{f\}$
         $G = G \setminus f$
      else
         $h = GG(G, w, \text{choose } c \in O_{el}(G))$
         $\delta = \delta \cup \{h\}$
      end if
      $G = G \setminus h$
   end if
end for
return $(\epsilon \cup \delta)$, $\Omega_{G,w}(\delta)$

---

of the paper in question. Generally speaking, TIGHT-CUT*, when compared to aforementioned, performed favorably (weighted instances included), and in certain number of instances with a huge advantage. TIGHT-CUT* makes it possible to compute solutions with tight approximation ratios within minutes, for instances that can be solved by exact methods only in hours or days. This result extends to real-world instances as well.

References

1. Achterberg, T.: SCIP: solving constraint integer programs. Math. Program. Comput. **1**(1), 1–41 (2009)
2. Ailon, N.: Aggregation of partial rankings, p-ratings and top-m lists. Algorithmica **57**(2), 284–300 (2008)
3. Ailon, N., Charikar, M., Newman, A.: Aggregating inconsistent information: ranking and clustering. J. ACM **55**(5), 1–27 (2008)

4. Alon, N., Lokshtanov, D., Saurabh, S.: Fast FAST. In: ICALP: International Colloquium on Automata, Languages, and Programming. Lecture Notes in Computer Science, vol. 5555, pp. 49–58. Springer, Berlin, Heidelberg (2009)
5. Alon, N., Spencer, J.: The Probabilistic Method. Wiley, New York (1992)
6. Ambainis, A., Balodis, K., Iraids, J., Kokainis, M., Prusis, K., Vihrovs, J.: Quantum speedups for exponential-time dynamic programming algorithms. In: SODA '19: Proceedings of the Thirtieth Annual ACM-SIAM Symposium on Discrete Algorithms, pp. 1783–1793. Society for Industrial and Applied Mathematics (2019)
7. Arditti, D.: A new algorithm for searching for an order induced by pairwise comparisons. In: E.D. et al. (ed.) Data Analysis and Informatics III, pp. 323–343. North Holland, Amsterdam (1984)
8. Ariyoshi, H., Higashiyama, Y.: A heuristic algorithm for the minimum feedback arc set problem. Res. Inst. Math. Anal. **427**, 112–130 (1981). Kyoto University Research Information Repository (Departmental Bulletin Paper)
9. Arora, S., Frieze, A., Kaplan, H.: A new rounding procedure for the assignment problem with applications to dense graph arrangement problems. Math. Program. **92**(1), 1–36 (2002)
10. Baharev, A., Schichl, H., Neumaier, A., Achterberg, T.: An exact method for the minimum feedback arc set problem. ACM J. Exp. Algorithm. **26**, 1–28 (2021)
11. Bang-Jensen, J., Gutin, G.: Digraphs: Theory, Algorithms and Applications. Springer. London (2002)
12. Bang-Jensen, J., Maddaloni, A., Saurabh, S.: Algorithms and kernels for feedback set problems in generalizations of tournaments. Algorithmica **76**(2), 320–343 (2015)
13. Bar-Yehuda, R.: One for the price of two: a unified approach for approximating covering problems. Algorithmica **27**(2), 131–144 (2000)
14. Barthelemy, J., Guenoche, A., Hudry, O.: Median linear orders: heuristics and a branch and bound algorithm. Eur. J. Oper. Res. **42**(3), 313–325 (1989)
15. Bartholdi, J., Tovey, C.A., Trick, M.A.: Voting schemes for which it can be difficult to tell who won the election. Soc. Choice Welfare **6**(2), 157–165 (1989)
16. Berger, B., Shor, P.W.: Approximation algorithms for the maximum acyclic subgraph problem. In: SODA '90: Proceedings of the First Annual ACM-SIAM Symposium on Discrete Algorithms, pp. 236–243. Society for Industrial and Applied Mathematics (1990)
17. Bessy, S., Fomin, F.V., Gaspers, S., Paul, C., Perez, A., Saurabh, S., Thomassé, S.: Kernels for feedback arc set in tournaments. J. Comput. Syst. Sci. **77**(6), 1071–1078 (2011)
18. Bhat, K.V., Kinariwala, B.: Optimum tearing in large scale systems and minimum feedback cutsets of a digraph. J. Franklin Inst. **307**(2), 83–94 (1979)
19. Biegler, L.T., Grossmann, I.E., Westerberg, A.W.: Systematic Methods for Chemical Process Design. Prentice Hall PTR (1997)
20. Bodlaender, H.L., Fomin, F.V., Koster, A.M.C.A., Kratsch, D., Thilikos, D.M.: A note on exact algorithms for vertex ordering problems on graphs. Theory Comput. Syst. **50**(3), 420–432 (2012)
21. Brandenburg, F.J., Hanauer, K.: Sorting heuristics for the feedback arc set problem—technical report mip-1104. Tech. rep., Department of Informatics and Mathematics, University of Passau, Germany (2011)
22. Brglez, F., Bryan, D., Kozminski, K.: Combinational profiles of sequential benchmark circuits. In: IEEE International Symposium on Circuits and Systems, pp. 1929–1934. IEEE, Piscataway (1989)
23. Bron, C.: Merge sort algorithm [m1]. Commun. ACM **15**(5), 357–358 (1972)
24. Burkard, R.E., Derigs, U.: Assignment and matching problems: solution methods with FORTRAN-programs. LN in Economics and Mathematical Systems, vol. 184. Springer, Berlin (1980)
25. Chanas, S., Kobylański, P.: A New Heuristic Algorithm Solving the Linear Ordering Problem. Comput. Optim. Appl. **6**(2), 191–205 (1996)
26. Charon, I., Guénoche, A., Hudry, O., Woirgard, F.: New results on the computation of median orders. Discrete Math. **165–166**, 139–153 (1997)

27. Charon, I., Hudry, O.: The noising method: a new method for combinatorial optimization. Oper. Res. Lett. **14**(3), 133–137 (1993)
28. Charon, I., Hudry, O.: A branch-and-bound algorithm to solve the linear ordering problem for weighted tournaments. Discrete Appl. Math. **154**(15), 2097–2116 (2006)
29. Cheung, L.K., Kuh, E.: The bordered triangular matrix and minimum essential sets of a digraph. IEEE Trans. Circuits Syst. **21**(5), 633–639 (1974)
30. Chvatal, V.: A greedy heuristic for the set-covering problem. Math. Oper. Res. **4**(3), 233–235 (1979)
31. Coleman, T., Wirth, A.: Ranking tournaments: Local search and a new algorithm. ACM J. Exp. Algorithm. **14**(2.6), 1–22 (2009)
32. Cook, S.A.: A taxonomy of problems with fast parallel algorithms. Inform. Control **64**(1–3), 2–22 (1985)
33. Coppersmith, D., Fleischer, L.K., Rurda, A.: Ordering by weighted number of wins gives a good ranking for weighted tournaments. ACM Trans. Algorithms **6**(3), 1–13 (2010)
34. Corbett, P.F.: Rotator graphs: an efficient topology for point-to-point multiprocessor networks. IEEE Trans. Parallel Distrib. Syst. **3**(5), 622–626 (1992)
35. Cormen, T.H., Leiserson, C.E., Rivest, R.L., Stein, C.: Introduction to Algorithms. MIT Press, Cambridge, MA (2009)
36. Csardi, G., Nepusz, T.: The igraph software package for complex network research (2006)
37. de Souza, C.C., Keunings, R., Wolsey, L.A., Zone, O.: A new approach to minimising the frontwidth in finite element calculations. Comput. Methods Appl. Mech. Eng. **111**(3–4), 323–334 (1994)
38. Decani, J.S.: A branch and bound algorithm for maximum likelihood paired comparison ranking. Biometrika **59**(1), 131–135 (1972)
39. Demetrescu, C., Finocchi, I.: Combinatorial algorithms for feedback problems in directed graphs. Inform. Proc. Lett. **86**(3), 129–136 (2003)
40. Diaz, M., Richard, J., Courvoisier, M.: A note on minimal and quasi-minimal essential sets in complex directed graphs. IEEE Trans. Circuit Theory **19**(5), 512–513 (1972)
41. Dom, M., Guo, J., Huffner, F., Niedermeier, R., Truss, A.: Fixed-parameter tractability results for feedback set problems in tournaments. In: CIAC: Italian Conference on Algorithms and Complexity, Lecture Notes in Computer Science, vol. 3998, pp. 320–331. Springer, Berlin-Heidelberg (2006)
42. Dom, M., Guo, J., Huffner, F., Niedermeier, R., Truss, A.: Fixed-parameter tractability results for feedback set problems in tournaments. J. Discrete Algorithms **8**(1), 76–86 (2010)
43. Dorigo, M., Gambardella, L.M.: Ant colony system: a cooperative learning approach to the traveling salesman problem. IEEE Trans. Evol. Comput. **1**(1), 53–66 (1997)
44. Du, D.Z., Hwang, F.K.: Generalized de Bruijn digraphs. Networks **18**(1), 27–38 (1988)
45. Dwork, C., Kumar, R., Naor, M., Sivakumar, D.: Rank aggregation methods for the web. In: WWW '01: Proceedings of the 10th international conference on World Wide Web, pp. 613–622. Association for Computing Machinery (2001)
46. Eades, P., Lin, X.: A heuristic for the feedback arc set problem. Aust. J. Comb. **12**, 15–25 (1995)
47. Eades, P., Lin, X., Smyth, W.: A fast and effective heuristic for the feedback arc set problem. Inform. Proc. Lett. **47**(6), 319–323 (1993)
48. Edwards, C.S.: A branch and bound algorithm for the Koopmans–Beckmann quadratic assignment problem. In: Mathematical Programming Studies. Mathematical Programming Studies, vol. 13, pp. 35–52. Springer, Berlin-Heidelberg (1980)
49. Even, G., Naor, J.S., Rao, S., Schieber, B.: Divide-and-conquer approximation algorithms via spreading metrics. J. ACM **47**(4), 585–616 (2000)
50. Even, G., Naor, J.S., Schieber, B., Sudan, M.: Approximating minimum feedback sets and multicuts in directed graphs. Algorithmica **20**(2), 151–174 (1998)
51. Feige, U.: Faster FAST (Feedback Arc Set in Tournaments) (2009)
52. Festa, P., Pardalos, P.M., Resende, M.G.C.: Feedback set problems. In: Handbook of Combinatorial Optimization, vol. A, pp. 209–258. Springer (1999)

53. Festa, P., Pardalos, P.M., Resende, M.G.C.: Algorithm 815: Fortran subroutines for computing approximate solutions of feedback set problems using grasp. ACM Trans. Math. Softw. **27**(4), 456–464 (2001)

54. Flood, M.M.: Exact and heuristic algorithms for the weighted feedback arc set problem: a special case of the skew-symmetric quadratic assignment problem. Networks **20**(1), 1–23 (1990)

55. Flueck, J.A., Korsh, J.F.: A branch search algorithm for maximum likelihood paired comparison ranking. Biometrika **61**(3), 621–626 (1974)

56. Fomin, F.V., Lokshtanov, D., Raman, V., Saurabh, S.: Fast local search algorithm for weighted feedback arc set in tournaments. In: AAAI'10: Proceedings of the Twenty-Fourth AAAI Conference on Artificial Intelligence, vol. 24, pp. 65–70. AAAI Press (2010)

57. Frieze, A., Kannan, R.: Quick approximation to matrices and applications. Combinatorica **19**(2), 175–220 (1999)

58. Galinier, P., Lemamou, E., Bouzidi, M.W.: Applying local search to the feedback vertex set problem. J. Heuristics **19**(5), 797–818 (2013)

59. Garey, M.R., Johnson, D.S.: Computers and Intractability: A Guide to the Theory of NP-Completeness. W. H. Freeman, San Francisco (1979)

60. Gillman, D.: A chernoff bound for random walks on expander graphs. SIAM J. Comput. **27**(4), 1203–1220 (1998)

61. Grotschel, M., Junger, M., Reinelt, G.: Acyclic subdigraphs and linear orderings: polytopes, facets, and a cutting plane algorithm. In: Graphs and Order. NATO ASI Series (Series C: Mathematical and Physical Sciences), vol. 147, pp. 217–264. Springer, Dordrecht (1985)

62. Grover, L.K.: A fast quantum mechanical algorithm for database search. In: Proceedings of the Twenty-Eighth Annual ACM Symposium on Theory of Computing—STOC '96, pp. 212–219. ACM Press (1996)

63. Guardabassi, G., Sangiovanni-Vincentelli, A.: A two levels algorithm for tearing. IEEE Trans. Circuits Syst. **23**(12), 783–791 (1976)

64. Gupta, S.: Feedback arc set problem in bipartite tournaments. Inform. Proc. Lett. **105**(4), 150–154 (2008)

65. Gupte, M., Shankar, P., Li, J., Muthukrishnan, S., Iftode, L.: Finding hierarchy in directed online social networks. In: Proceedings of the 20th International Conference on World Wide Web, pp. 557–566. ACM, New York (2011)

66. Gurevich, Y., Shelah, S.: Expected computation time for Hamiltonian path problem. SIAM J. Comput. **16**(3), 486–502 (1987)

67. Gurobi Optimization, LLC.: Gurobi Optimizer

68. Hagberg, A.A., Schult, D.A., Swart, P.J.: Exploring network structure, dynamics, and function using networkx. In: In Proceedings of the 7th Python in Science Conference (SciPy2008), pp. 11–15 (2008)

69. Hassin, R., Rubinstein, S.: Approximations for the maximum acyclic subgraph problem. Inform. Proc. Lett. **51**(3), 133–140 (1994)

70. Hecht, M., Gonciarz, K., Horvát, S.: Tight localizations of feedback sets. ACM J. Exp. Algorithmics **26**, 1–19 (2021)

71. Hecht, M.S., Ullman, J.D.: Flow graph reducibility. SIAM J. Comput. **1**(2), 188–202 (1972)

72. Hecht, M.S., Ullman, J.D.: Characterizations of reducible flow graphs. J. ACM **21**(3), 367–375 (1974)

73. Hoare, C.A.R.: Algorithm 64: quicksort. Commun. ACM **4**(7), 321 (1961)

74. Horvat, S.: IGraph/M (2020).

75. Itoh, M: A design for directed graphs with minimum diameter. IEEE Trans. Comput. **C-32**(8), 782–784 (1983)

76. Impagliazzo, R., Paturi, R., Zane, F.: Which problems have strongly exponential complexity? J. Comput. Syst. Sci. **63**(4), 512–530 (2001)

77. Johnson, D.B.: Finding all the elementary circuits of a directed graph. SIAM J. Comput. **4**(1), 77–84 (1975)

78. Kaas, R.: A branch and bound algorithm for the acyclic subgraph problem. Eur. J. Oper. Res. **8**(4), 355–362 (1981)
79. Kaku, B.K., Thompson, G.L., Baybars, I.: A heuristic method for the multi-story layout problem. Eur. J. Oper. Res. **37**(3), 384–397 (1988)
80. Karpinski, M., Schudy, W.: Faster algorithms for feedback arc set tournament, Kemeny rank aggregation and betweenness tournament. In: Algorithms and Computation. Lecture Notes in Computer Science, vol. 6506, pp. 3–14. Springer, Berlin (2010)
81. Kautz, W.H.: Unit-distance error-checking codes. IEEE Trans. Electron. Comput. **EC-7**(2), 179–180 (1958)
82. Kendall, M.G.: Further contributions to the theory of paired comparisons. Biometrics **11**(1), 43 (1955)
83. Kenyon-Mathieu, C., Schudy, W.: How to rank with few errors. In: Proceedings of the Thirty-Ninth Annual ACM Symposium on Theory of Computing—STOC '07, pp. 95–103. ACM Press (2007)
84. Klein, P., Stein, C., Tardos, É.: Leighton-rao might be practical: faster approximation algorithms for concurrent flow with uniform capacities. In: Proceedings of the Twenty-Second Annual ACM Symposium on Theory of Computing—STOC '90, pp. 310–321. ACM Press (1990)
85. Knuth, D.E.: Sorting and Searching. The Art of Computer Programming, vol. 3, 2nd edn. Addison-Wesley Professional, Reading, MA (1998)
86. Koehler, H.: A contraction algorithm for finding minimal feedback sets. In: ACSC '05: Proceedings of the Twenty-eighth Australasian Conference on Computer Science, vol. 38, pp. 165–173. Australian Computer Society (2005)
87. Kudelić, R.: Monte-carlo randomized algorithm for minimum feedback arc set. Appl. Soft Comput. **41**, 235–246 (2016)
88. Kudelić, R., Ivković, N.: Ant inspired monte carlo algorithm for minimum feedback arc set. Expert Syst. Appl. **122**, 108–117 (2019)
89. Kudelić, R., Rabuzin, K.: Dealing with intractability of information system subsystems development order via control flow graph reducibility. In: Proceedings of the 2020 3rd International Conference on Electronics and Electrical Engineering Technology, pp. 62–68. ACM, New York (2020)
90. Kuo, C.J., Hsu, C.C., Lin, H.R., Chen, D.R.: Minimum feedback arc sets in rotator graphs. In: The 26th Workshop on Combinatorial Mathematics and Computation Theory, pp. 95–101 (2009)
91. Lee, W., Rudd, D.F.: On the ordering of recycle calculations. AIChE J. **12**(6), 1184–1190 (1966)
92. Leighton, T., Rao, S.: An approximate max-flow min-cut theorem for uniform multicommodity flow problems with applications to approximation algorithms. In: [Proceedings 1988] 29th Annual Symposium on Foundations of Computer Science, pp. 422–431. IEEE, Piscataway (1988)
93. Lempel, A., Cederbaum, I.: Minimum feedback arc and vertex sets of a directed graph. IEEE Trans. Circuit Theory **13**(4), 399–403 (1966)
94. Levy, H., Low, D.W.: A contraction algorithm for finding small cycle cutsets. J. Algorithms **9**(4), 470–493 (1988)
95. Lubotzky, A., Phillips, R., Sarnak, P.: Ramanujan graphs. Combinatorica **8**(3), 261–277 (1988)
96. Luby, M., Nisan, N.: A parallel approximation algorithm for positive linear programming. In: Proceedings of the Twenty-Fifth Annual ACM Symposium on Theory of Computing—STOC '93, pp. 448–457. ACM Press (1993)
97. Mathieu, C., Schudy, W.: How to Rank with Fewer Errors: A PTAS for Feedback Arc Set in Tournaments. Preliminary version in STOC 2007 (Kenyon-Mathieu and Schudy 2007)
98. Mathieu, C., Schudy, W.: Yet another algorithm for dense max cut: go greedy. In: SODA '08: Proceedings of the nineteenth annual ACM-SIAM symposium on Discrete algorithms, pp. 176–182. Society for Industrial and Applied Mathematics (2008)

99. Matuszewski, C., Schonfeld, R., Molitor, P.: Using sifting for k-layer straightline crossing minimization. In: Graph Drawing. Lecture Notes in Computer Science, vol. 1731, pp. 217–224. Springer, Berlin (1999)

100. Mezard, M., Montanari, A.: Information, Physics, and Computation. Oxford University Press, Oxford (2009)

101. Muñoz, X., Unger, W., Vrt'o, I.: One sided crossing minimization is NP-hard for sparse graphs. In: GD: International Symposium on Graph Drawing. Lecture Notes in Computer Science, vol. 2265, pp. 115–123. Springer, Berlin (2002)

102. Noughabi, H.A., Baghbani, F.G.: An efficient genetic algorithm for the feedback set problems. In: 2014 Iranian Conference on Intelligent Systems (ICIS). IEEE, Piscataway (2014)

103. Eades, P., Smyth, W.F., Lin, X.: Heuristics for the Feedback Arc Set Problem. Techreport 1, School of Computing Science, Curtin University of Technology, Perth, Western Australia (1989)

104. Pardalos, P.M., Qian, T., Resende, M.G.: A greedy randomized adaptive search procedure for the feedback vertex set problem. J. Comb. Optim. 2(4), 399–412 (1998)

105. Park, S., Akers, S.: An efficient method for finding a minimal feedback arc set in directed graphs. In: [Proceedings] 1992 IEEE International Symposium on Circuits and Systems, pp. 1863–1866. IEEE, Piscataway (1992)

106. Phillips, J.P.N.: A Procedure for determining Slater's i and all nearest adjoining orders. Br. J. Math. Stat. Psychol. 20(2), 217–225 (1967)

107. Phillips, J.P.N.: A Further procedure for determining Slater's i and all nearest adjoining orders. Br. J. Math. Stat. Psychol. 22(1), 97–101 (1969)

108. Pho, T.K., Lapidus, L.: Topics in computer-aided design: part I. An optimum tearing algorithm for recycle systems. AIChE J. 19(6), 1170–1181 (1973)

109. Raghavan, P.: Probabilistic construction of deterministic algorithms: approximating packing integer programs. J. Comput. Syst. Sci. 37(2), 130–143 (1988)

110. Ramachandran, V.: Finding a minimum feedback arc set in reducible flow graphs. J. Algorithms 9(3), 299–313 (1988)

111. Ramachandran, V.: A minimax arc theorem for reducible flow graphs. SIAM J. Discrete Math. 3(4), 554–560 (1990)

112. Raman, V., Saurabh, S.: Improved parameterized algorithms for feedback set problems in weighted tournaments. In: Parameterized and Exact Computation. Lecture Notes in Computer Science, vol. 3162, pp. 260–270. Springer, Berlin (2004)

113. Raman, V., Saurabh, S.: Parameterized algorithms for feedback set problems and their duals in tournaments. Theor. Comput. Sci. 351(3), 446–458 (2006)

114. Raman, V., Saurabh, S., Sikdar, S.: Improved exact exponential algorithms for vertex bipartization and other problems. In: ICTCS: Italian Conference on Theoretical Computer Science. LNCS, vol. 3701, pp. 375–389. Springer, Berlin (2005)

115. Raman, V., Saurabh, S., Sikdar, S.: Efficient exact algorithms through enumerating maximal independent sets and other techniques. Theory Comput. Syst. 41(3), 563–587 (2007)

116. Rosen, B.K.: Robust linear algorithms for cutsets. J. Algorithms 3(3), 205–217 (1982)

117. Saab, Y.: A fast and effective algorithm for the feedback arc set problem. J. Heuristics 7(3), 235–250 (2001)

118. Saab, Y.G.: A fast and robust network bisection algorithm. IEEE Trans. Comput. 44(7), 903–913 (1995)

119. Schwikowski, B., Speckenmeyer, E.: On enumerating all minimal solutions of feedback problems. Discrete Appl. Math. 117(1-3), 253–265 (2002)

120. Sedgewick, R., Wayne, K.: Algorithms, 4th edn. Addison-Wesley, Reading, MA (2011)

121. Seymour, P.D.: Packing directed circuits fractionally. Combinatorica 15(2), 281–288 (1995)

122. Shamir, A.: A linear time algorithm for finding minimum cutsets in reducible graphs. SIAM J. Comput. 8(4), 645–655 (1979)

123. Simpson, M., Srinivasan, V., Thomo, A.: Efficient computation of feedback arc set at web-scale. Proc. VLDB Endowment 10(3), 133–144 (2016)

124. Smith, G., Walford, R.: The identification of a minimal feedback vertex set of a directed graph. IEEE Trans. Circuits Syst. **22**(1), 9–15 (1975)
125. Spencer, J.: Ten Lectures on the Probabilistic Method, 2nd edn. Society for Industrial and Applied Mathematics (1994). (1987 edition not acquirable)
126. Tarjan, R.: Depth-first search and linear graph algorithms. SIAM J. Comput. **1**(2), 146–160 (1972)
127. Tarjan, R.E.: Testing flow graph reducibility. J. Comput. Syst. Sci. **9**(3), 355–365 (1974)
128. Tucker, A.W.: On directed graphs and integer programs. In: Symposium on Combinatorial Problems. Princeton University, Princeton (1960)
129. Waggener, W.M.: Pulse Code Modulation Techniques, 1st edn. Springer (1995)
130. Wang, C.C., Lloyd, E.L., Soffa, M.L.: Feedback vertex sets and cyclically reducible graphs. J. ACM **32**(2), 296–313 (1985)
131. Xiao, H.: Packing feedback arc sets in reducible flow graphs. J. Comb. Optim. **32**(3), 951–959 (2015)
132. Xu, Y.Z., Zhou, H.J.: Optimal segmentation of directed graph and the minimum number of feedback arcs. J. Stat. Phys. **169**(1), 187–202 (2017)
133. Younger, D.: Minimum feedback arc sets for a directed graph. IEEE Trans. Circuit Theory **10**(2), 238–245 (1963)
134. Zhao, J.H., Zhou, H.J.: Optimal Disruption of Complex Networks (2016)
135. van Zuylen, A.: Linear programming based approximation algorithms for feedback set problems in bipartite tournaments. Theor. Comput. Sci. **412**(23), 2556–2561 (2011)
136. van Zuylen, A., Hegde, R., Jain, K., Williamson, D.P.: Deterministic pivoting algorithms for constrained ranking and clustering problems. In: SODA '07: Proceedings of the Eighteenth Annual ACM-SIAM Symposium on Discrete Algorithms, pp. 405–414. Society for Industrial and Applied Mathematics (2007)
137. van Zuylen, A., Williamson, D.P.: Deterministic algorithms for rank aggregation and other ranking and clustering problems. In: Approximation and Online Algorithms. LNCS, vol. 4927, pp. 260–273. Springer, Berlin (2007)
138. van Zuylen, A., Williamson, D.P.: Deterministic pivoting algorithms for constrained ranking and clustering problems. Math. Oper. Res. **34**(3), 594–620 (2009)

Part III
Complexity Informed

Chapter 4
Having the Right Tool

4.1 Tackling With Feedback Arc Set

While FAS is indeed quite difficult problem to tackle, there are some good news as well. An exact method[1] for MFAS can be found in [2], the method is based on minimum Set Cover and enumeration of all simple cycles[2]—please consult Sect. 3.42 of the book.

The problem is also fixed-parameter tractable, namely it is possible to solve it in polynomial time, in $O(n^4 4^k k^3 k!)$ to be exact, if k is bounded above by a constant, via $FVS \preceq FAS$ linear time [7] reducibility [5].

On general graphs, FAS can be approximated to within $O(\log(|V|) \log(\log(|V|)))$ [7]—for details one should consult Sect. 3.17 of the book. There is also possibility of approximating FAS in polynomial time with arbitrary probability via Monte-Carlo approach from [16] (upgraded ACO inspired version is presented in [17])—Sects. 3.43 and 3.48 of the book.

On tournament, PTAS exists and can be found in [15], for $\epsilon > 0$, outputted ranking expected cost is at most $(1 + \epsilon)OPT$—Sect. 3.27 of the book. There is also parameterized and sub-exponential algorithm for weighted version of FAST, presented in [14]—Sect. 3.36 of the book.

General heuristic for FAS can be found in [12], where empirical validation has shown that approximation is $r \leq 2$—Sect. 3.49 of the book. For a web-scale application, with millions of nodes and billions of arcs, one should consult [24], specifically [6] from Eades et. al. (Sect. 3.13 of the book) and [4] from Berger et.

---

[1] There is also exact algorithm found in [11], that uses Dynamic Programming and yields solution in $O(2^m |E|^4 \log(|V|))$, where $m \leq |E| - |V| + 1$ (m can be computed in $O(|E|^3)$ time).

[2] In cases encountered during research, only a tractable number of cycles had to be enumerated until MFAS was found.

© The Author(s), under exclusive license to Springer Nature Switzerland AG 2022
R. Kudelić, *Feedback Arc Set*, SpringerBriefs in Computer Science,
https://doi.org/10.1007/978-3-031-10515-9_4

al. (Sect. 3.9 of the book) as well (which were two best performers of web-scale research, respectively)—Sect. 3.45 of the book.

For a quantum approach, that is based on Grover's search [10], the answer is research that can be found in [1]. This paper has provided quantum algorithm for directed FAS which runs in $O^*(1.728^n)$ time [1]—Sect. 3.47 of the book.

There are also other good algorithms not mentioned in this short guide, therefore the reader is advised to consult Sect. 3 of the book (Papers and Algorithms), and as necessary, online supplementary material where one can find additional resources.[3]

4.2 Tractability in Special Cases

There are instances of FAS where tractability is possible, and if one can reduce his own instance of a problem being tackled to some of these special cases, then an optimal solution is efficiently acquirable.

In a situation where an input digraph is constrained to a planar graph, we have a problem instance where MFAS can be solved in polynomial time [3, 18]. If an input graph is Reducible Flow Graph (RFG), then the problem of MFAS is also efficiently solvable in polynomial time [19, 22, 23]—Sect. 3.7 of the book.

Furthermore, there is also a class of graphs called weakly acyclic digraphs for which MFAS can be solved in polynomial time (considering that planar digraphs are also weakly acyclic this is not surprising—Sect. 3.6 of the book) [9], and if we are dealing with a $K_{3,3}$-free digraph ("a digraph such that its underlying graph does not contain as a subgraph $K_{3,3}$, or any of its subdivisions"), then this also represents a special case where MFAS is polynomially solvable [21].

It is also possible to solve MFAS efficiently for a rotator graph, that is in linear time, in terms of the number of arcs in the rotator graph, nevertheless it should be noted that an n-rotator graph has $n!$ vertices [20]—Sect. 3.30 of the book.

There are also times when we have a graph whose edges are undirected,[4] at such an instance a problem of deleting a number of edges with whose removal this graph would be made acyclic is easily solvable in polynomial time by finding a Minimum Spanning Tree (MST) of such a graph [8, 13].

---

[3] Mentioned in the preface of this book.

[4] On the other hand it is interesting to note that Feedback Vertex Set problem remains NP-Complete even for undirected graph [12], but is solvable in polynomial time "with no vertex degree exceeding 3 by reducing to the Matroid Parity problem" [25].

References

1. Ambainis, A., Balodis, K., Iraids, J., Kokainis, M., Prusis, K., Vihrovs, J.: Quantum speedups for exponential-time dynamic programming algorithms. In: SODA '19: Proceedings of the Thirtieth Annual ACM-SIAM Symposium on Discrete Algorithms, pp. 1783–1793. Society for Industrial and Applied Mathematics (2019)
2. Baharev, A., Schichl, H., Neumaier, A., Achterberg, T.: An exact method for the minimum feedback arc set problem. ACM J. Exp. Algorithm. **26**, 1–28 (2021)
3. Bang-Jensen, J., Gutin, G.: Digraphs: Theory, Algorithms and Applications. Springer. London (2002)
4. Berger, B., Shor, P.W.: Approximation algorithms for the maximum acyclic subgraph problem. In: SODA '90: Proceedings of the First Annual ACM-SIAM Symposium on Discrete Algorithms, pp. 236–243. Society for Industrial and Applied Mathematics (1990)
5. Chen, J., Liu, Y., Lu, S., O'sullivan, B., Razgon, I.: A fixed-parameter algorithm for the directed feedback vertex set problem. J. ACM **55**(5), 1–19 (2008)
6. Eades, P., Lin, X., Smyth, W.: A fast and effective heuristic for the feedback arc set problem. Inform. Proc. Lett. **47**(6), 319–323 (1993)
7. Even, G., Naor, J.S., Schieber, B., Sudan, M.: Approximating minimum feedback sets and multicuts in directed graphs. Algorithmica **20**(2), 151–174 (1998)
8. Gabow, H.N., Galil, Z., Spencer, T., Tarjan, R.E.: Efficient algorithms for finding minimum spanning trees in undirected and directed graphs. Combinatorica **6**(2), 109–122 (1986)
9. Grotschel, M., Junger, M., Reinelt, G.: Acyclic subdigraphs and linear orderings: polytopes, facets, and a cutting plane algorithm. In: Graphs and Order. NATO ASI Series (Series C: Mathematical and Physical Sciences), vol. 147, pp. 217–264. Springer, Dordrecht (1985)
10. Grover, L.K.: A fast quantum mechanical algorithm for database search. In: Proceedings of the Twenty-Eighth Annual ACM Symposium on Theory of Computing—STOC '96, pp. 212–219. ACM Press (1996)
11. Hecht, M.: Exact Localisations of Feedback Sets. Theory Comput. Syst. **62**(5), 1048–1084 (2017)
12. Hecht, M., Gonciarz, K., Horvát, S.: Tight localizations of feedback sets. ACM J. Exp. Algorithmics **26**, 1–19 (2021)
13. Karger, D.R., Klein, P.N., Tarjan, R.E.: A randomized linear-time algorithm to find minimum spanning trees. J. ACM **42**(2), 321–328 (1995)
14. Karpinski, M., Schudy, W.: Faster algorithms for feedback arc set tournament, Kemeny rank aggregation and betweenness tournament. In: Algorithms and Computation. Lecture Notes in Computer Science, vol. 6506, pp. 3–14. Springer, Berlin (2010)
15. Kenyon-Mathieu, C., Schudy, W.: How to rank with few errors. In: Proceedings of the Thirty-Ninth Annual ACM Symposium on Theory of Computing—STOC '07, pp. 95–103. ACM Press (2007)
16. Kudelić, R.: Monte-carlo randomized algorithm for minimum feedback arc set. Appl. Soft Comput. **41**, 235–246 (2016)
17. Kudelić, R., Ivković, N.: Ant inspired monte carlo algorithm for minimum feedback arc set. Expert Syst. Appl. **122**, 108–117 (2019)
18. Kudelić, R., Konecki, M.: On the Minimum Feedback Arc Set: Planarity of information systems digraphs. In: Proceedings of The 11th MAC 2017, pp. 155–160. MAC Prague Consulting s.r.o. (2017)
19. Kudelić, R., Rabuzin, K.: Dealing with intractability of information system subsystems development order via control flow graph reducibility. In: Proceedings of the 2020 3rd International Conference on Electronics and Electrical Engineering Technology, pp. 62–68. ACM, New York (2020)
20. Kuo, C.J., Hsu, C.C., Lin, H.R., Chen, D.R.: Minimum feedback arc sets in rotator graphs. In: The 26th Workshop on Combinatorial Mathematics and Computation Theory, pp. 95–101 (2009)

21. Nutov, Z., Penn, M.: On the integral dicycle packings and covers and the linear ordering polytope. Discrete Appl. Math. **60**(1-3), 293–309 (1995)
22. Ramachandran, V.: Finding a minimum feedback arc set in reducible flow graphs. J. Algorithms **9**(3), 299–313 (1988)
23. Ramachandran, V.: A minimax arc theorem for reducible flow graphs. SIAM J. Discrete Math. **3**(4), 554–560 (1990)
24. Simpson, M., Srinivasan, V., Thomo, A.: Efficient computation of feedback arc set at web-scale. Proc. VLDB Endowment **10**(3), 133–144 (2016)
25. Ueno, S., Kajitani, Y., Gotoh, S.: On the nonseparating independent set problem and feedback set problem for graphs with no vertex degree exceeding three. Discrete Math. **72**(1–3), 355–360 (1988)

Printed in the United States
by Baker & Taylor Publisher Services.

Printed in the United States
by Baker & Taylor Publisher Services